Bernd Ulmann
AN/FSQ-7: the computer that shaped the Cold War

Christmas 2022

Bernd Ulmann

AN/FSQ-7:
the computer that shaped the Cold War

Author
Prof. Dr. Bernd Ulmann
65307 Bad Schwalbach
Germany
ulmann@vaxman.de

ISBN 978-3-486-72766-1
e-ISBN 978-3-486-85670-5
Set-ISBN 978-3-486-85671-2

Library of Congress Cataloging-in-Publication Data
A CIP catalog record for this book has been applied for at the Library of Congress.

Bibliographic information published by the German National Library
The German National Library lists this publication in the Deutsche Nationalbibliografie; detailed bibliographic data are available on the Internet at http://dnb.dnb.de.

© 2014 Oldenbourg Wissenschaftsverlag GmbH
Rosenheimer Straße 143, 81671 München, Germany
www.degruyter.com
Part of De Gruyter

Editor: Dr. Gerhard Pappert
Production editor: Tina Bonertz
Cover picture: Ron Brunell
Print and Binding: CPI books GmbH, Leck

Printed in Germany
This paper is resistant to aging (DIN/ISO 9706).

To my beloved wife Rikka.

Acknowledgments

This book would not have been possible without the support and help of many people. First of all, I would like to thank my wife Rikka Mitsam who never complained about being neglected for many months while I lived more or less in seclusion, reading memoranda, reports and manuals, dissecting some of the few AN/FSQ-7 programming examples and writing this book. In addition to this she did a grand job in researching the whereabouts of the various Direction and Control Centers which culminated in table 4.1. Furthermore she worked miracles with some of the pictures in this book which were blurry or had to be drawn altogether by hand.

I am particularly indebted to Jens Breitenbach who did a magnificent job at proof reading and who provided a plethora of suggestions and improvements which enhanced this book significantly. He also spotted and corrected many of my LaTeX-sins thus also improving the overall appearance of this book.

In addition to that I would like to thank Daniela Koch for her invaluable support at proof reading and her many suggestions which made the book much more readable. Dirk Bartkowiak, Arne Charlet, and Patrick Hedfeld also spotted many errors and made suggestions which improved the text considerably.

Furthermore, I am indebted to Sara Lott, Dag Spicer, and Al Kossow from the Computer History Museum in Mountain View (http://www.computerhistory.org/). Without the document collection at http://www.bitsavers.org this book would not have been possible at all – accordingly, I would like to thank all persons who contributed to this incredible collection of historical documents.

Joachim E. Wagner worked miracles with some of the pictures which were next to unusable in their original state.

Many people who worked on and with SAGE and the AN/FSQ-7 supported me in a variety of ways by sharing their recollections, scans of documents, and answering many questions, detailed and general alike. Thus I would like to thank the following persons (in alphabetical order): David E. Casteel, Captain, USAF (ret) (who also did a lot of proofreading), David Fankhauser, Mike Loewen and Glen Williamson. Last, but not least, I would like to thank Marcin Wichary and Ron Brunell for their permission to use some of their pictures (figures 9.8 and 5.38). I am also indebted to Mr. Brunell for valuable information about the FETRONs mentioned in section 5.16. He also contributed the cover photo.

This book was typeset with LaTeX, the vector graphics were created with xfig and most of the schematics were drawn with EAGLE.

Registered names, trademarks, designations etc. used in this book, even when not specifically marked as such, are not to be considered unprotected by law.

Contents

1	Introduction	1
2	**Setting the stage**	3
2.1	Computers until 1945	3
2.2	Basics of vacuum tube circuits	8
2.3	Toward Whirlwind and beyond	12
3	**Whirlwind**	21
3.1	Basic architecture	21
3.1.1	Arithmetic element	27
3.1.2	Control	31
3.2	Storage	33
3.2.1	The Massachusetts Institute of Technology (MIT) Storage Tube	36
3.2.2	Magnetic core storage	40
3.3	Magnetic drums	47
3.4	Magnetic tapes	51
3.5	Paper tape readers, punches and typewriters	53
3.6	Data transmission	54
3.7	Oscilloscope displays and light guns	56
3.8	Time register	58
3.9	Reliability, power supplies and marginal checking	58
3.10	Programming	62
3.11	The end of Whirlwind	66
4	**SAGE**	69
4.1	The Cape Cod system	71
4.2	SAGE and AN/FSQ-7	72
4.3	Control and direction centers	75
4.4	AN/FSQ-7 overview	81

5	**Basic circuitry**	**89**
5.1	Introduction	89
5.2	Cathode follower	92
5.3	Pulse amplifier	93
5.4	Register driver	94
5.5	Relay drivers	96
5.6	Level setter	98
5.7	Diode AND and OR circuits	100
5.8	Gate tube circuit	101
5.9	DC inverter	102
5.10	Flip-flops	103
5.11	Single-shots	106
5.12	Pulse generators	109
5.13	Delay lines and delay line drivers	111
5.14	Special circuits	112
5.15	Pluggable units	113
5.16	The FETRON	115
5.17	Troubleshooting	116
6	**Central processor**	**119**
6.1	Memory element	122
6.2	Instruction control element	126
6.3	Selection and IO control element	128
6.4	Program element	129
6.5	Arithmetic element	131
7	**Drum system**	**139**
7.1	Magnetic drums	140
7.2	Timing	144
7.3	Status concept and time stamps	146
7.4	Data flows	146

8	**Input/output system**	**149**
8.1	Input system	149
8.1.1	Long-Range Input (LRI) element	150
8.1.2	Gap-Filler Input (GFI) element	153
8.1.3	Crosstell (XTL) element	156
8.1.4	Test Pattern Generator (TPG)	157
8.2	Manual data input element	157
8.3	Output system	158
8.4	Alarms and warning lights	162
8.5	Tape drives and card machines	163
9	**Display system**	**165**
9.1	Situation display	167
9.2	Light gun	171
9.3	Area discriminators	173
9.4	Digital display	174
9.5	Photographic recorder-reproducer element	176
10	**Machine consoles**	**179**
11	**Power supply**	**183**
11.1	The powerhouse	183
11.2	Regulated power supplies	184
11.3	Power distribution	186
11.4	Marginal checking system	188
12	**Programming**	**191**
12.1	Instruction format	191
12.2	Instruction set	192
12.2.1	Miscellaneous class	193
12.2.2	Add class	194
12.2.3	Multiply class	195
12.2.4	Store class	196
12.2.5	Shift class	197
12.2.6	Branch class	198
12.2.7	Input/output class	198
12.2.8	Reset class	199
12.3	Indexed addressing	199

12.4	Subroutines	200
12.5	Examples	201
12.5.1	Polynomial evaluation	202
12.5.2	Coordinate transformation	202
12.5.3	Finding the largest number	204
12.5.4	Adding ten numbers	205
12.5.5	Delaying	206
12.5.6	Printing	207
12.5.7	Trick programs	207

13 Software — 209

13.1	Software development process	209
13.2	Operational software	212

14 Failure or Success? — 215

14.1	A failure?	215
14.2	Success!	217
14.2.1	Hardware	217
14.2.2	Graphics	218
14.2.3	Software	219
14.2.4	Air Traffic Control (ATC) and SABRE	220
14.2.5	Semi Automatic Ground Environment (SAGE) in popular culture	220

15 Epilogue — 221

A Whirlwind instruction set — 223

B Programming cards — 229

Bibliography — 233

Acronyms — 247

Index — 251

1 Introduction

Many books covering various aspects of the *Cold War*[1] have been written in the past – most of them deal with the highly complex chains of events which took place after World War II and during the Cold War, many deal with personal histories and recollections, often heroic and brave, but only few concentrate on the machines which were developed during this time as a direct result of new requirements posed by the unique threats of the Cold War. Many of these machines would eventually turn out to be cornerstones of our technological age, ranging from high-speed aircraft, rockets, manned space-flight to miniaturized electronics, high-speed digital computers and much more. The effect the Cold War had on the development of computer systems cannot be overestimated. According to [EDWARDS 1997][p. 61] about 75 to 80 percent of the total revenue of the US American computer industry in 1950 was due to developments requested by the military. Accordingly he considers the effect of the Cold War on the development of digital computers as the most important asset of the Cold War:

> "Of all the technologies built to fight the Cold War, digital computers have become its most ubiquitous, and perhaps its most important, legacy."[2]

In contrast to many other books, this book mainly focuses on technical and technological aspects of two particular machines that not only shaped the Cold War as such but also pioneered many of the the technologies we take for granted today. These two machines are Whirlwind which is covered in some detail in chapter 3 and its successor, the stored-program digital computer at the very heart of the *AN/FSQ-7*[3] *Combat Direction Central* and the *AN/FSQ-8 Combat Control Central*.[4]

Reportedly, 56 of these behemoth computer systems were built, most of them configured as a so-called *duplex* system consisting of a pair of AN/FSQ-7 (or AN/FSQ-8) computers. The rationale behind coupling two of these machines in a duplex configuration was to minimize downtime – one machine normally acting as the *active*

---

[1] The so-called *Cold War* describes the often bordering on the hot-side struggle between the United States of America and the Soviet Union. This struggle went on for 43 years from 1946 to 1989 (see [Legacy 2000][p. 2-1]). One of the most important aspects of the Cold War from a technological point of view is the fact that money was not an issue at all as long as no side had gained a stable advantage.

[2] See [EDWARDS 1997][p. ix]. Other authors like HEINZ ZEMANEK obviously thought along the same lines (see [ZEMANEK 1977][p. 7]): *"Wer an den Fortschritt glaubt, muß an den Computer glauben, erstens grundsätzlich und zweitens praktisch, denn neben Raumfahrt und Atomkraft ist der Computer die einzig spektakuläre Nachkriegsneuheit am Fortschritt."* This can be roughly translated as follows: "Believing in progress is believing in the computer – in a fundamental as well as a practical way since the computer is the only spectacular achievement after World War II except for space-flight and nuclear power."

[3] Short for *Army-Navy Fixed Special eQuipment*, colloquially just known as the Q7.

[4] Since the AN/FSQ-8 is architecturally quite similar to the AN/FSQ-7 but lacks much of the latter input and output equipment, the following text will focus on the AN/FSQ-7.

computer, the other as *standby*. 27 of these duplex systems, each weighing in excess of 250 tons, containing more than 50,000 vacuum tubes and requiring about three megawatts of electrical energy for a typical installation, were eventually built.[5] Some ran for nearly a quarter of a century. The network consisting of these vast computer installations and associated input/output equipment comprised what became known as *SAGE*, the *Semi-Automatic Ground Environment*[6] which implemented the *Air Defense (AD)* of North America. Although largely forgotten and neglected, these systems paved the way for many of our today's technologies, ranging from real-time processing and operating systems, data transmission over telephone lines, the foundations of software engineering to time-sharing, graphical user interfaces, and much more.

Quite a lot of books and articles have been written about SAGE and the AN/FSQ-7 covering nearly all perspectives ranging from a nationwide view, trhough those of the institutions which were part of the massive development efforts, to personal recollections.[7] Interestingly yet no book focuses on the machine itself, which was so incredibly ahead of its time that it still seems modern in many respects when looking back from today's perspective. This is the gap this book tries to fill: It is dedicated to this very special computer and its predecessor *Whirlwind*, both of which will be described in detail with the focus on technical and technological aspects.

But before focusing on these two computers, the following chapter describes the state of the art of computing at the end of World War II, which was dominated by analog computers, most of which still were intricate electromechanical devices with only a few forays into the domain of digital computing. It also contains a short description of the events and decisions which led to the development of Whirlwind and its final application to the *air defense problem*.[8]

The following chapter 3 is then devoted exclusively to Whirlwind, the predecessor of the AN/FSQ-7 computer which paved the way for core memory, real-time operation, graphical displays, high reliability, and many more.

Chapter 4 focuses on the SAGE system as such, describing the military and technical environment into which the AN/FSQ-7 had to fit. The remaining chapters then describe the hardware and programming model of AN/FSQ-7 in detail. Chapter 14 will focus on SAGE-trivia, the unavoidable question whether SAGE and the AN/FSQ-7 were a failure or even a fraud as some people have claimed, and describes the most important legacies of SAGE and some of its more or less direct spin-offs. Finally, chapter 15 ends this book with some remarks on the influence of the Cold War on pure research.

---

[5] Except for one site which was located in Canada, those installations were spread over the United States.

[6] Some malicious tongues insisted that SAGE was the acronym for *"Someone Always Gets Excited"*.

[7] [REDMOND et al. 1975], [REDMOND et al. 1980], [REDMOND et al. 2000], [MURPHY 1972] and [JACOBS 1986] are worth reading.

[8] The intricate interdependencies of the various institutes, laboratories etc. which developed Whirlwind and later the AN/FSQ-7 and the overall SAGE system will be covered only in such detail as necessary to understand the origins of the various technological developments. See [REDMOND et al. 1975], [REDMOND et al. 1980], [REDMOND et al. 2000], [GREEN 2010], [EDWARDS 1997], [JACOBS 1986], [GROMETSTEIN 2011], [MITRE 2008] and [MURPHY 1972] for more information on these topics.

2 Setting the stage

It is worthwhile to have a short look[1] at the state of the art of computing at the end of World War II. Most of these computing devices were analog in nature with only a few digital computers, which were still considered being inferior to analog computers.[2] Most of these developments were the direct result of military requirements like fire-control systems, the computation of firing tables etc.

The technologies available at the end of this war and the experiences gathered by engineers and scientists who devoted their knowledge to the application of automatic computing devices to all sorts of problems, would lead to an unprecedented flourishing of electronics in general and computers in particular. The ideas invented during the years of war were so fruitful that they would inspire a whole generation of engineers and scientists. Eventually, the result of their work would be the ubiquitous computing devices on which our everyday lives depend so heavily. The most important and most neglected of these computers are those machines most people never see: Control systems, ranging from small embedded computers which control simple machinery like a household appliance to heavily interconnected systems with global influence and importance. These systems take care of the consumables we need, they control the flight of our airplanes, they direct the traffic in our cities etc. Without these computers our civilization would invariably fail to exist and most of these devices owe to early developments after World War II, the most notable of those being Whirlwind and AN/FSQ-7.

2.1 Computers until 1945

Of all the various computing systems developed during World War II, most machines were so-called *analog computers*,[3] which are based on the principle of implementing a model of a problem to be solved. Early such analog computers were purely mechanical devices like VANNEVAR BUSH's[4] *differential analyzers*, the first of which was completed in 1931 at the *Massachusetts Institute of Technology, MIT* for short. Similar mechanical

---

[1] A comprehensive history of computing in these early years is far beyond the scope of this book, so only a couple of examples will be accounted for.

[2] In fact, analog computers were superior devices when compared with stored-program digital computers for quite some time, but the development of Whirlwind showed for the first time that the realm of real-time applications and integrated control systems was not only that of analog computers, although it was only in the 1970s that digital computers were able to beat analog computers in terms of speed, ease of application and real-time ability, finally replacing them altogether.

[3] See [ULMANN 2013].

[4] 03/11/1890–06/28/1974

devices were built by Douglas Rayner Hartree[5] and Arthur Porter[6] at Manchester University in 1934, by a group led by Svein Rosseland[7] and many more. As powerful as these machines were, from a mathematical point of view, they were big, heavy and programming was a time consuming task. Programming back then required one to connect the various computing elements like summers, integrators, multipliers and the like with gears and rods, a task that often required hours and sometimes days to perform. This was alleviated a bit with the development of electromechanical differential analyzers. These still used mechanical computing elements but relied on intricate electrical servo-systems to interconnect the computing elements. Now the task of programming only required rewiring of a central patch-panel.

While these general-purpose analog computers were mostly one-of-a-kind machines, there was high demand for special-purpose analog computers which were employed aboard ships, submarines and on land for tasks like automatic range keeping and fire-control.[8]

Already during World War II it became clear that the days of these large, heavy and intricate mechanisms which required labor-intensive maintenance on a regular schedule were numbered. Electronic analog computers were about to take over. As so often, when the time for an idea has come, electronic analog computers were developed independently in the United States of America by George Arthur Philbrick[9] and in Germany by Helmut Hoelzer.[10] While Philbrick's machine, called *Polyphemus*, due to its appearance (a rack with a single oscilloscope mounted on the top while the analog computer itself occupied the lower part), was aimed at process-simulation, the machines developed by Hoelzer had their roots in the war-time development of the *A4*.[11]

From the perspective of this book, Hoelzer's developments are especially interesting, since they not only produced radically novel devices which advanced the state of the art of their time but they also were the result of a war-time development. He not only developed the first truly general-purpose electronic analog computer,[12] which by the way was used well after the war during the development of the *Redstone* and *Jupiter* rocket as well as for the design of the first satellite of the United States, *Explorer I*.[13] In addition to this, he developed the so-called *Mischgerät*[14] which was the world's first fully electronic on-board flight computer that controlled the flight of the A4 rocket.[15]

---

[5]03/27/1897–02/12/1958, see [Fischer 2004].

[6]12/08/1910–02/26/2010

[7]03/18/1894–01/19/1985

[8]Such devices were already developed and used during World War I – see [Friedman 2008] for more information about this topic, later fire-control computers are also treated in [Svoboda 1948] and [Ulmann 2013][pp. 21 ff. and pp. 47 ff.].

[9]01/05/1913–12/01/1974

[10]02/27/1912–08/19/1996

[11]Short for *Aggregat 4*, the name used during its development. The resulting weapon system became known as *V2*, short for *Vergeltungswaffe 2*.

[12]See [Tomayko 1985].

[13]See [Tomayko 2000][p. 236].

[14]This can be roughly translated as *mixing unit*.

[15]See [Lange 2006] or [Ulmann 2013][p. 31 ff.].

2.1 Computers until 1945

Only a few digital computers were actually developed before and during World War II. One of the early pioneers of digital computers was KONRAD ZUSE[16] who built the first binary, program-controlled, although not stored-program, computer, known as the Z3 which was followed by the quite successful Z4. This latter machine was not destroyed during the war as the Z3 and was used successfully for quite some years at the ETH Zürich in Switzerland.[17]

Another development of ZUSE during World War II were two special-purpose computers, the *S1* and *S2*.[18] These machines were intended to automatically compute corrections for wings of glide bombs like the *Hs 293*. One especially remarkable feature of the more sophisticated S2 was the existence of an analog-digital-converter which allowed the system to compute values based on measurements transmitted directly to the computer. Due to constraints caused by the ongoing war, all of ZUSE's machines were based on relays as those found in telephone exchanges, although the idea of using vacuum tubes as the active elements in his computers had already received quite some thought.

Other digital computer developments were due to JOHN VINCENT ATANASOFF[19] and CLIFFORD EDWARD BERRY[20] who developed the ATANASOFF-BERRY *Computer, ABC* for short, beginning in the late 1930s. This machine was completed in 1942.[21] It used a binary representation of data but was not truly programmable – in fact, the ABC was a special-purpose machine[22] aimed at the solution of systems of linear equations.

A much larger scale development in the field of programmable digital computers was done under the auspices of HOWARD HATHAWAY AIKEN[23] and led to the *Automatic Sequence Controlled Calculator, ASCC* for short.[24] This system had breathtaking dimensions compared with other developments like those described above. It weighed about 4.5 tons and measured 51 feet in length, eight feet in height and two feet in depth and was entirely electromechanical.[25]

In 1943 JOHN WILLIAM MAUCHLY[26] and JOHN ADAM PRESPER ECKERT[27] began the development of *ENIAC*, the *Electronic Numerical Integrator and Computer*, shown in figure 2.1, which was originally intended to calculate artillery firing tables.[28] Eventually ENIAC, which was put into operation on February 14th, 1946, turned out to be the largest fully electronic digital computer in its day, containing about 18,000 vacuum tubes, 1,500 relays, 7,200 semiconductor diodes and tens of thousands of capacitors

---

[16] 06/22/1910–12/18/1995
[17] See [BRUDERER 2012] for more information on this.
[18] See [ZUSE 1993][p. 62].
[19] 10/04/1903–06/15/1995
[20] 04/19/1918–10/30/1963
[21] See [GUSTAFSON 2000] for more information.
[22] Although there is some debate concerning this term, see [GUSTAFSON 2000][pp. 103 f.].
[23] 03/03/1900–03/14/1973
[24] This machine was renamed *Mark I* and is also known as *Harvard Mark I*.
[25] See [COHEN 2000] for more in-depth information.
[26] 04/30/1907–01/08/1980
[27] 04/09/1919–06/03/1995
[28] ENIAC is often mentioned as the "first" digital computer, but there is considerably controversy about this notion due to developments like those described above.

Figure 2.1: U.S. Army photo of ENIAC (initial installation at the Moore School of Electrical Engineering, it was transferred to the Ballistic Research Laboratories, Aberdeen Proving Ground, in 1947), cpl. IRWIN GOLDSTEIN in the foreground sets switches on one of ENIAC's function tables.[31]

and resistors. The overall machine weighed about 27 tons and consumed 140 kW.[29] Although naysayers doubted that a machine using so many vacuum tubes could perform any useful work at all between the presumably short periods between two failures, ENIAC proved them wrong.[30] The basis for this success was a conservative circuit design in conjunction with some simple operating procedures like not turning the computer on and off but let it run to reduce thermal stress on the tubes.

Although ENIAC came too late to calculate firing tables for artillery shells it was put into use for a variety of mainly military projects, the first of which was suggested as early as 1945 by JOHN VON NEUMANN:[32]

---

[29] See [EDWARDS 1997][p. 50]. [VAN DER SPIEGEL et al. 2000] gives an architectural overview of the machine and describes a modern implementation on a VLSI chip. A rather contemporary description of ENIAC may be found in [STIFLER et al. 1950][pp. 194 ff.].

[30] Those naysayers based their assumptions on experiences with the most complex electronic devices prior to ENIAC which contained several hundred and in some cases even more than 1,000 vacuum tubes.

[31] Picture source: http://commons.wikimedia.org/wiki/File:Classic_shot_of_the_ENIAC_%28full_resolution%29.jpg, retrieved 11/27/2013.

[32] JÁNOS LAJOS NEUMANN, 12/28/1903–02/08/1957

> "In early 1945, as the construction of the ENIAC was nearing completion, VON NEUMANN raised the question with [physicist STANLEY] FRANKEL and METROPOLIS of using it to perform the very complex calculations involved in hydrogen bomb design. The response was immediate and enthusiastic. Arrangements were made by VON NEUMANN on the basis that the 'Los Alamos problem' would provide a much more severe challenge to the ENIAC on its shakedown trial."[33]

These calculations proved to be quite successful which in 1946 led the director of Los Alamos to write a letter to the Moore School stating that *"the complexity of these problems is so great that it would have been impossible to arrive at any solution without the aid of ENIAC."*[34] In fact, ENIAC proved to be such a valuable tool that the press was enthusiastic and created terms like *"giant brain"* and others. A typical report from the ENIAC unveiling read like this:

> *"Mathematical brain enlarges man's horizons... A new epoch in the history of human thought began last night. The scope and area in which man's brain can grasp, predic[t], control suddenly opened outward into the distance with revelation of section construction during the war of a 30 ton mathematical brain that solves the unsolvable."*[35]

Other headlines were equally enthusiastic and exaggerating. The time-honored London Times featured an article about ENIAC with the headline *"An Electronic Brain: Solving Abstruse Problems; Valves with a Memory"* – too much anthropomorphism for DOUGLAS RAYNER HARTREE,[36] a renowned British mathematician and physicist. He sent a sharply-worded letter to the editor but to no avail. The notion of the giant or electronic brain persisted.[37]

It turned out that ENIAC's architecture, which had few things in common with today's computer architectures and more closely resembled a so-called *Digital Differential Analyzer*,[38] a digital variant of an analog computer, was not suitable for future developments, but more importantly, this machine proved that computers involving tens of thousands of delicate vacuum tubes were not the maintenance nightmare that had been predicted. Although not comparable with today's machines in terms of reliability and *mean time to repair*, *MTTR* for short, it was reliable enough to allow the solution of problems which were next to unthinkable before. In this respect ENIAC

---

[33] See [RHODES 2005][p. 251] and [MARTIN 1995].

[34] See [EDWARDS 1997][p. 51].

[35] See [ECKERT 1986][p. 3].

[36] 03/27/1897–02/12/1958

[37] The diction of the *electronic* or *giant brain* was so powerful, that [FORRESTER et al. 1948][p. 5] explicitly noted that *"Despite its popular reputation, the digital computer does not have human thinking intelligence, but it does have superhuman computing speed, and, when properly directed (viz., fed input information correctly and programmed correctly to describe the desired operations), the output gives orders of information which can be either automatically transformed into action or interpreted by humans for further processing."* It also may have added much to the eventual disappointment of the early and strong artificial intelligence approaches which could not fulfill the dreams of machines capable of thought.

[38] *DDA* for short.

deserves highest respect for encouraging not only scientists and engineers to employ highly complex electronic computers but also for being the archetypal *giant brain* that would shape the picture of a computer for the following years.

2.2 Basics of vacuum tube circuits

Since many of today's readers may not be familiar with vacuum tube circuits any more, a short digression might be allowed since Whirlwind and AN/FSQ-7 both were vacuum tube digital computers. Some basic knowledge about vacuum tubes, these nearly forgotten active circuit elements, is helpful for the following chapters.

Although there was a plethora of vacuum tubes in use until the late 1960s when transistors finally won the battle as *the* all-purpose active element, we will focus only on diodes and triodes to get an impression of how basic vacuum tube circuits work.[39] Basically the simplest vacuum tube, the *diode* consists of a heated *cathode* which emits electrons and an *anode* which collects those electrons.

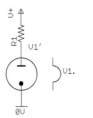

Figure 2.2: Basic vacuum diode circuit

Figure 2.2 shows the schematic of a simple circuit employing a vacuum diode V1. The cathode is typically made from nickel and is coated with an alkaline earth metal oxide which gives off electrons rather easily. It is depicted by the dot connected to circuit ground while the anode, which is also often called *plate*, is represented by the plate-like symbol connected to the resistor R1. The symbol to the right of the tube itself denotes the *heater* which heats the cathode to 800–1000° C. Such an arrangement is called an *indirectly heated cathode* – simpler tubes use the emission from the glowing heater itself and are called *directly heated*. Indirectly heated vacuum tubes have some advantages, especially the fact that heater and cathode are essentially isolated against each other. The heater won't be shown in following schematics.

If the anode is at a positive potential with regard to the cathode, as shown in figure 2.2, electrons will flow from cathode to anode and the tube is said to be *conducting*, i. e. a current flows through the resistor R1.[40] If the anode is negative with respect to the cathode, its potential will repel the electrons and no current flows. Thus a vacuum diode allows current to flow in one direction only – quite like a modern semiconductor diode.[41]

A far more interesting device is the *triode* which is based on a vacuum diode which has been extended by a so-called *control grid* which normally is just a fine wire mesh placed between the electron emitting cathode and the anode. If this grid is left unconnected,

---

[39] Although pentodes were also used often in early computer circuits, these will not be covered in this section.

[40] Typical anode or plate voltages are in the range of a couple hundred volts.

[41] This comparison is rather simplified but suffices for the purposes of this book. More detailed information about tubes, their characteristics and use can be found in one of the classics of vacuum tube technology like [VALLEY 1948] or [BARRY 1949]. [HUSKEY 2000] gives a nice introduction into basic vacuum circuits suitable for digital computers.

2.2 Basics of vacuum tube circuits

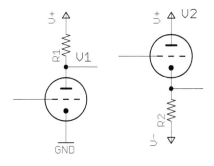

Figure 2.4: Basic vacuum triode circuits, inverter left and cathode follower right

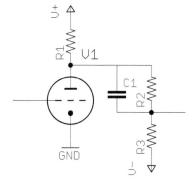

Figure 2.5: Inverter with output network

the tube just behaves like a vacuum diode, but applying a more or less negative voltage to this grid allows for modulation of the electron flow from the cathode to the anode, hence the name *control* grid. The interesting thing about this means of control is that a triode is controlled by a voltage, not a current as would be the case in a bipolar transistor. Thus a vacuum triode may be compared to a depletion field-effect transistor.

Figure 2.3 shows a typical so-called *twin-triode* of the 1940s, i. e. two triodes sharing a common evacuated glass envelope. In the middle of the front triode the fine mesh comprising the grid can be seen, surrounding the thin and long indirectly heated cathode which in turn contains the heater windings. The grid in turn is surrounded by two halves of the anode. Two such triodes are housed next to each other. At the bottom the wires connecting heaters, cathodes, grids and anodes to the socket connectors protrude through the glass envelope.

Figure 2.4 shows two basic triode circuits. In both cases, the triode and a resistor form a controlled voltage divider. The circuit on the left is a so-called *inverter*: If the grid is at ground potential or slightly above,[42] current flows from cathode to anode, so the junction between the plate resistor R1 and the plate will be pulled to ground potential. Since the tube and the resistor form a voltage divider, the output potential at the plate-resistor junction will be rather low with respect to ground. If the grid voltage is made negative with respect to the cathode, the flow of electrons will be blocked to a certain degree which depends on the grid and plate voltage and the tube's overall characteristics.

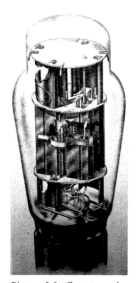

Figure 2.3: Cutaway view of a twin-triode 6AS7G (see [BARRY 1949][p. 551])

---

[42]The control grid of a tube should normally be at a small positive potential with respect to the cathode at maximum. Otherwise it would no longer just control the flow of the electrons but attract them instead of the anode which would cause a considerable current to flow through the grid, which in turn would heat the grid, overload the vacuum tube and eventually destroy it.

So a negative voltage at the control grid causes the output of this left circuit to be positive while a zero or slightly positive grid voltage will cause the output to go low which makes this circuit behave like an inverter. If the control grid were connected to a number of different signal sources, each decoupled by a semiconductor diode, a basic digital gate like a *NAND*, short for *Not And*, will result.

The circuit shown on the right in figure 2.4 is a so-called *cathode follower*. Here the resistor and the tube have switched places, so that a conducting tube (non-negative control grid voltage) will cause a positive output while a negative grid voltage will cause a negative output. So, basically, the cathode follower is a non-inverting circuit and can be used readily to amplify a digital signal.

In practice these two circuits are much too simplified for use in high-speed computer circuits. One of the main problems are stray capacitances in the tube itself and its wiring especially with respect to the output lines. Controlling the grid of either the inverter or the cathode follower with a square wave signal will realistically result in an output signal which deviates more or less from a square wave, since the rising and falling edges will be deformed according to an exponential curve due to those stray capacitances. The basic solution to this problem is shown in figure 2.5 which resembles an inverter with the addition of an output network consisting of two resistors R2 and R3 in series where one resistor, R2, is paralleled by a capacitor C1. When the polarity of the output signal changes, a rather large current flows through C1 thereby counteracting the deformation of the rising or falling edges caused by the unavoidable stray capacitances. Given the large physical dimensions of vacuum tube based computers where signal lines often have a length in meters rather than centimeters, these capacitances of the output lines would make high-speed operation impossible without such countermeasures.

Figure 2.6 shows one of the most fundamental digital circuits of all, a *flip-flop*. This is a bistable circuit which can be used to store one bit or to build counters etc. It basically consists of two inverters connected in a crosswise fashion: If the left tube conducts, the potential at its anode which is connected to the control grid of the right triode via R3, is at a negative potential through R6 so the right tube does not conduct. Applying a positive impulse to the grid of the right tube then lets this tube conduct, which in turn blocks the left triode which will then hold the right tube in the conducting state.

Figure 2.7 shows a typical early application of such a triode flip-flop: A ring-counter circuit from ENIAC. In contrast to today's computers, ENIAC was a decimal instead of a binary machine. As if this number representation was not unusual enough, ENIAC used such ring-counters consisting of ten stages to represent a single decimal digit, so a single *accumulator*[43] capable of storing one ten-digit number consisted of ten such flip-flops, each requiring two triodes or one twin-triode.[44]

This ring-counter consists of ten flip-flop stages similar to that shown in figure 2.6. Each stage is coupled crosswise between the grids and anodes by means of an

---

[43] An accumulator is an internal register of a computer which is normally connected quite directly to the arithmetic logic unit.

[44] [STIFLER et al. 1950][p. 25] points out that in addition to these 200 triodes for a single accumulator about 400 extra tubes were necessary for implementing carry-handling, reset, signal shaping etc.

2.2 Basics of vacuum tube circuits

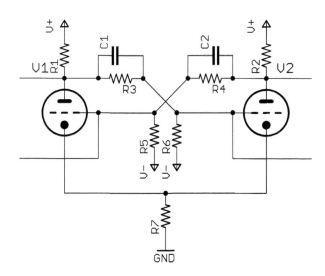

Figure 2.6: Basic flip-flop circuit with two triodes

RC-combination[45] and is in one of two stable states at each time. These ten flip-flop stages are in turn coupled by capacitors connecting the anode of the right triode of a pair with the control grid of the right anode of the flip-flop right next to it.

The shading denotes the active triodes – the leftmost flip-flop is in the *set* state while the remaining nine flip-flops are all *reset*. It is interesting to note how this circuit is set to a particular value: Its input is shown on the right – incoming negative pulses are applied to the cathodes of the left triodes of each flip-flop element. A negative pulse will only reset the one particular flip-flop which was in the set state. This reset-operation results in a pulse transmitted through the interstage coupling capacitor to the next flip-flop in the chain which will in turn be set.

It is clear that such a number-representation requires much more circuit elements per digit than even a decimal based binary representation like the so-called *BCD*-code.[46] This representation uses ten out of the 16 possible states of four bits. So storing a single BCD digit would have required only four flip-flops with some additional circuitry to avoid illegal states not representing a decimal digit. Nevertheless the savings would have been substantial. Even more savings would have been possible if a pure binary representation of values in the computer had been chosen from the beginning.

Due to habit and fear of radical changes like switching to a then uncommon number system, many early computers relied on the decimal number system yielding unnecessarily complex circuits as seen from today's perspective. It is interesting to note that KONRAD ZUSE not only decided to use a pure binary representation of values at the

---

[45] A resistor with capacitor wired in parallel.
[46] Short for *binary coded decimal*.

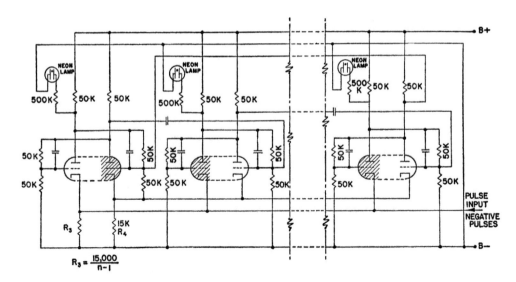

Figure 2.7: Typical ring-counter circuit as used in ENIAC (see [STIFLER et al. 1950][p. 24])

very beginning of his developments in the 1930s[47] but also incorporated special circuits in his machines which performed the necessary conversions between the external decimal and the internal binary number representation.

2.3 Toward Whirlwind and beyond

Only military applications could offer the necessary amount of financial backing and staffing necessary to build a large scale stored-program digital computer back in the early days of computing. One particularly important and complex problem of World War II was the simulation of aircraft for research as well as for training purposes. Training new pilots in real aircraft not only bound many aircraft in the training centers instead of using them in battle, but was also a dangerous way to teach the art of flying. Not too surprisingly, many casualties among the pilots and even more damaged aircraft were the result, rendering the traditional way of training more than ineffective under the immense pressure of an ongoing war.

Intrigued by the problem of aircraft simulation,[48] LUIS DE FLOREZ[49] – characterized as being *"brilliant [and] flamboyant engineer"*[50] – asked his alma mater, the MIT, to build a flight simulator. This machine should be used for research purposes as well as

---

[47] In fact his machines were floating point processors.

[48] As history shows, the simulation of aircraft with a considerable degree of precision according to the effects taken into account, should become and stay one of the main applications of large electronic analog computer installations. Only in the 1970s digital computers became powerful enough to take over this crucial task.

[49] 03/04/1889–12/05/1962

[50] See [REDMOND et al. 1980][p. 1].

2.3 Toward Whirlwind and beyond

for training naval bomber flight crews without exposing them to the unnecessary risks of real flights. This request should finally lead to the development of *Whirlwind*,[51] sponsored by the *Office of Naval Research (ONR)*, which should not only become the icon of an era but extend the frontiers of electronic computing in an unprecedented way, being the first real-time digital computer.

When work on the proposed flight simulator began, it was decided to build an analog computer according to the then state of the art in electronic circuit design. This seemed a plausible decision since flight simulation was and still is one of the archetypal problems requiring real-time computation, something that was thought to be possible only by analog computing. Instead of building a training device, it was decided in about 1944 to focus on research on the stability and control characteristics of large aircraft in general by building an analog computer, the *Aircraft Stability and Control Analyzer (ASCA)*.

As early as in 1946, JAY WRIGHT FORRESTER[52] decided to abandon the analog computer approach to ASCA and start development of a novel stored-program digital computer,[53] which eventually would become the world's first *real-time* computer suitable for control and simulation tasks. As a direct result of this shift of focus, the work previously done at the *Servomechanisms Laboratory*[54] was relocated to a new laboratory, the *Digital Computer Laboratory (DCL)*, which was set up to take over the task of developing this new digital computer.[55] The DCL then operated as a division of the Servomechanisms Laboratory until it became an independent laboratory for digital computers and their applications.[56]

Let us digress a bit and have a look at the situation of air defense right after the end of World War II: Back then the United States operated more than 70 so-called *Ground Control Intercept (GCI)* sites which were rather intricate installations, each consisting of one or two *search radars*, *height-finder radars*, and various ground-to-air and air-to-ground communication facilities to direct intercept aircraft to approaching bombers and the like.[57]

First contacts between the *Air Defense Evaluation Group (ADSEC)* and the team working on Whirlwind were made as early as 1947.[58] When the Soviet Union successfully detonated their first nuclear weapon in 1949, it became clear that the United States were no longer the only global power armed with nuclear weapons and some action would be necessary to assure that a first-strike by the Soviet Union could be answered immediately to assure a climate of mutual deterrence. Thus in November 1949 GEORGE

---

[51] This rather menacing name should become the source of a pun when HEINZ ZEMANEK (*01/01/1920) who developed a transistorized computer in Austria in the 1950s named this particular machine "Mailüfterl" with the rationale that while his group in Austria would not be able to build "Typhoon" or a "Whirlwind", at least a light Viennese breeze in May might be accomplished.
[52] *07/14/1918
[53] This was partially due to an encounter of FORRESTER with PERRY ORSON CRAWFORD who proposed digital computation as early as 1939 in his Master's thesis (see [FORRESTER 1988] and [CRAWFORD 1939]).
[54] Originally established in 1940 to develop automatic control systems and military fire control.
[55] See [JACOBS 1986][pp. 8 f.].
[56] See [MIT 1955][p. 1].
[57] See [JACOBS 1986][p. 2].
[58] See [REDMOND et al. 1980][p. 32].

EDWARD VALLEY Jr.[59] from MIT proposed to THEODOR VON KÁRMÁN[60] that a general study of air defense requirements should be undertaken immediately to be prepared for any possible Soviet aggression.[61]

The results of this study turned out to be devastating. In 1950 the ADSEC under GEORGE E. VALLEY and the *Weapon Systems Evaluation Group (WSEG)* reported that the current air defense system of the United States, more or less a remnant of the systems in operation at the end of World War II, was no longer sufficient for current and future threats involving nuclear weapons and strategic long-range bombers capable of delivering such weapons. Rather drastically, the *Scientific Advisory Board (SAB)* remarked that the current manual system was in fact *"lame, purblind, and idiot-like"* with extra emphasis on the last point:

> *"[Of] these comparatives, idiotic is the strongest. It makes little sense for us to strengthen the muscles if there is no brain; and given a brain, it needs good eyesight."*[62]

It was clear that future requirements for such an air defense system would be substantially different from those encountered and mastered quite well by means of manual systems during World War II:

> *"The earth's curvature meant that hundreds, if not thousands, of radars would be required to detect low-flying aircraft... There was no conceivable way in which human radar operators could be employed to make [the necessary] calculations for hundreds of aircraft as detected from such a large number of radars, nor could the data be coordinated into a single map if the operators used voice communications. The [...] computations were straightforward enough. [...] It was doing all that work in real time that was impossible."*[63]

These findings led to a study of the overall task of air defense which was performed as *Project Charles* at the MIT. This project suggested that the existing manual system should be upgraded to fill the most pressing immediate needs while in parallel a laboratory which would become known as the *Lincoln Laboratory*[64] should be established to work on a so-called *transition system* which would eventually lead to the development of the AN/FSQ-7 and *SAGE*.

It is interesting to note that FORRESTER and some of his colleagues already proposed the application of a computer to the overall air defense task two years earlier, back in 1948, which pretty much antedates the findings of the ADSEC and WSEG and their conclusions:

---

[59] 09/05/1913–10/16/1999

[60] 05/11/1881–05/06/1963

[61] See [GROMETSTEIN 2011][pp. 1 ff.] and [JACOBS 1986][p. xv].

[62] See [REDMOND et al. 1980][pp. 172 f.].

[63] See [EDWARDS 1997][p. 91].

[64] The laboratory was to be built at the intersection of the towns of Bedford, Lexington and Lincoln; since there already were projects named "Bedford" and "Lexington" in operation, naming the new laboratory "Lincoln Laboratory" was an obvious choice (see [GROMETSTEIN 2011][p. 8]). [GROMETSTEIN 2011] provides an in-depth account of the Lincoln Laboratory's history.

2.3 Toward Whirlwind and beyond

> *"An example of a requirement in the solution of tactical problems is found in the interception of supersonic missiles. Here there is need for [a] combined computer and control system which can:*
>
> 1. *Automatically receive radar and other information from multiple locations,*
> 2. *Correlate this with past information,*
> 3. *Distinguish between types of missiles and distinguish missiles from aircraft based on identifying information and trajectories,*
> 4. *Predict trajectories to the impact point (if flight is uncontrolled),*
> 5. *Assess possible damage and importance of defense action to permit concentration of defense against the most dangerous missiles,*
> 6. *Take rapid automatic defensive action in selecting launching location and firing defensive missiles, where time is too short for human intervention,*
> 7. *Carry on these operations with a minimum of equipment,*
> 8. *Possess the required flexibility to avoid the need for redesigning to meet changing tactical situations and the appearance of new weapons of either offense or defense."*[65]

In January 1950 paths crossed, when GEORGE E. VALLEY learned from JEROME BERT WIESNER[66] – whom he met by chance in a hall of MIT – that there already was a computer named Whirlwind that was *"sitting up for grabs on the MIT campus"*.[67] This should be the beginning of one of the biggest adventures in digital computing and would finally lead to the development of the first large scale nationwide computer network for the purpose of air defense. At this time Whirlwind[68] was already operating partially while development of and on the machine was still ongoing.[69] Whirlwind was completed in stages with the overall system being fully operational in 1951.[70]

Already in 1950, an experimental system based on Whirlwind as its central component was set up to demonstrate the feasibility of the application of a large scale digital computer to the air defense problem.[71] Since Whirlwind was still in an early stage of development, only 256 memory cells of electrostatic memory and an additional 32 memory locations of so-called *test memory* were available.[72] Accordingly, the overall program to perform the *track-while-scan* operation with all necessary constants etc. had to fit into this tiny amount of memory, next to incredible from today's point of view.[73]

---

[65] See [FORRESTER et al. 1948][pp. 8 f.].
[66] 05/30/1915–10/21/1994
[67] See [JOHNSON 2002][p. 124].
[68] Block diagrams of the computer had already been developed and published in 1947, see [MIT 1955][p. 1].
[69] See [REDMOND et al. 1975][pp. 1.08 ff.], [REDMOND et al. 1980] and [GREEN 2010] for a thorough account of the events which led to Whirlwind's development.
[70] See [TCM 83][p. 13]. Also in 1951, most of the researchers working on the application of Whirlwind to the air defense problem at the DCL joined the Lincoln Laboratory.
[71] See [ISRAEL 1951] for more in-depth information about this demonstration, the setup, and the program developed.
[72] Some of these test memory locations were used for memory mapped IO.
[73] Dissecting the program listing given in [ISRAEL 1951][pp. 50 f.] is a very worthwhile task.

> *"The objective of the first air defense experiments and studies was the use of the Whirlwind Computer to perform the necessary computational and data-processing functions associated with:*
>
> > a *automatic track-while-scan (Track While Scan (TWS)); that is, the automatic tracking and display of selected aircraft using data obtained from a continuously-rotating search radar, and*
> >
> > b *the automatic track-while-scan of selected aircraft and the computation of the heading instructions necessary to guide one aircraft – the interceptor – on a collision course with a second aircraft – the target. These interception computations were to be such that they could be used for the mid-course phase of the interception, leading to a closing phase under the direction of airborne intercept (Airborne Intercept (AI)) radar."*[74]

Figure 2.8: DRR connecting the MEW radar to Whirlwind (see [WIESER 1983][p. 363])

To accomplish this task, Whirlwind – located at the MIT in Boston – was coupled to a *Microwave Early Warning (MEW)* radar set at Bedford airport, about 12 miles away from Boston. To transmit the radar data to Whirlwind, an already existing prototype *Digital Radar Relay (DRR)* was used.[75] The other communications link to the interceptor aircraft was implemented by means of a traditional voice radio link.

First experiments during the fall of 1950 were not successful due to problems with the DRR but in the spring of 1951 actual tests involving aircraft supplied by the Instrumentation Laboratory of the MIT and the *Air Force Cambridge Research Center (AFCRC)* commenced. These tests were highly successful as [ISRAEL 1951][p. 6] notes:[76]

> *"These tests established that the computer with a storage capacity of 256 registers*[77] *could successfully track five aircraft*[78] *or could track two aircraft, guiding one on a collision course interception with the other. About ten interception flight tests were attempted and completed through June of 1951; the final separations of the target and the interceptor aircraft as their paths crossed averaged between 500 and 1500 yards."*

---

[74] See [ISRAEL 1951][pp. 4 f.].
[75] See [HARRINGTON 1983] for an in-depth description of the development of radar data transmission.
[76] See also [WIESER 1983][p. 365].
[77] One such *register* corresponds to one memory location of 16 bits.
[78] Even a program tracking nine aircraft was developed and tested with *"some success"*.

2.3 Toward Whirlwind and beyond

Due to the restrictions of the MEW radar system, the analog-to-digital conversion equipment,[79] the DRR, and Whirlwind, some simplifications were necessary to perform these tests: First of all only two dimensions were taken into account, so target and interceptor were assumed to be acting in the same horizontal plane. Azimuth data was converted into an eight bit value, so an angular resolution of about 1.4° could be achieved. Range data was converted into a seven bit value so values up to 127 miles could be represented. This range information was padded with a 0 bit yielding an eight bit byte, too. For data transmission the azimuth byte was padded with two bits 01 while the range data was padded with 00. Transmission of one such ten bit word took 1/50 of a second.

On the receiving end these ten bits were stored in a 16 bit wide so-called *Flip-Flop Register* which could be addressed as a normal memory location.[80] Writing into this register was synchronized with the operation of the computer to ensure data integrity by blocking write accesses while the computer reads this particular register. Since Whirlwind did not feature interrupts, a polling scheme was employed which checked this register more frequent than every 1/50th of a second for new data.[81]

During these studies it became clear that multiplication operations were executed much more often than divisions which was due to the fact that the data transmitted by the radar system via the DRR naturally represented polar coordinates while all computations took place in Cartesian coordinates due to the Cartesian nature of the display system, so a lot of coordinate transformations had to be performed. This observation would have a great impact on the design of the arithmetic unit of the later AN/FSQ-7 since an implicit shift was built into its addition operation to speed up multiplications. The detrimental effect this had on divisions was accepted and programmers tried to avoid divisions when- and wherever possible.

Nearly all of the experiments based on the Bedford radar station used so-called *manual acquisition* in which a human operator had to select a particular target aircraft to be tracked. This acquisition process was implemented by a light gun, which mainly consisted of a photomultiplier tube built into a gun-like enclosure so that an operator could point the tube's window to a particular area of interest on the display screen. When the object displayed in that area of the display was updated the next time by the computer, an impulse was generated by the light gun's circuitry which reset a bit in a Flip-Flop register. The computer then had to check this particular bit after drawing every single object to determine which object the light gun was pointed to when its trigger was pulled.

These early experiments also led to the development of the *radar mapper*: These devices were used to remove clutter which was especially pronounced at near distances. A *Plan Position Indicator (PPI)* display showed incoming radar data; above this display, a photomultiplier tube with the necessary optics was mounted so that it could

---

[79]This analog-to-digital conversion equipment was based on two simple counters which were reset regularly, corresponding to azimuth and range being zero respectively (see [HARRINGTON 1983][p. 371]).

[80]Today, this would be called *memory mapped IO*.

[81]After reading a value from this Flip-Flop register, a negative zero, i.e. 11...111, was written to it. Since the first pad bit of either azimuth or range data was zero, new data could be easily checked for by a simple comparison operation.

Register length	16 binary digits, parallel
Speed	20,000 single-address operations per second
Storage capacity	Originally 256 registers
	Recently 320 registers
	Presently 1,280 registers
	Target 2,048 registers
Order type	Single-address, one order per word
Numbers	Fixed point
Basic pulse repetition frequency	1 megacycle
	2 megacycles (arithmetic element only)
Tube count	5,000, mostly single pentodes
Crystal count	11,000

Table 2.1: Characteristics of Whirlwind as of 1951 (see [EVERETT 1951][p. 71])

"see" the overall display. This tube registered only the blue flash of light which was emitted when the display was updated. To remove clutter, those areas of the display tube where static features like land marks etc. were displayed, could be covered by an opaque mask which was painted manually onto the display tube's surface. Thus the photomultiplier assembly only generated an output signal when an update on an area of the PPI was performed which was not masked. This signal was then used to gate data transmission to the receiving computer, effectively suppressing any clutter signals.[82]

To get an impression of the amount of computing power available at the time of these early experiments, table 2.1 shows the characteristics of Whirlwind as of 1951. Apart from the scarce amount of memory available, another aspect is remarkable: That of the rather short word length. Other contemporary machines, which were mainly aimed at scientific and engineering computations, featured word lengths of 36 or 40 bits while Whirlwind only used a 16 bit machine word. The rationale behind this decision was that it was preferable and easy to trade machine time for precision by using subroutines to perform multi-precision computations instead of building a much larger system involving much more delicate components like vacuum tubes that were error prone and would not only complicate the computer unnecessarily but turn the machine into a maintenance nightmare. FORRESTER remembers:

> "Making the decision to build Whirlwind [...] with a 16 binary digit register length was tremendously hard for us. The mathematicians were up in arms. They thought it was too short to be of any possible use. We defended it at that time on the basis that it was a demonstration of feasibility and we would build a 32 or a 36 bit computer when the right time came. [...] Selecting 16 bits was not a useless register length for computing, only a serious short term political problem."[83]

---

[82]See [WIESER 1983][p. 367].
[83]See [FORRESTER 1988][p. 13].

2.3 Toward Whirlwind and beyond

STORAGE (16-digit words)

Magnetic-Core Storage (Magnetic-Core Storage (MS)):
2 banks, each having 1,024 cores per digital plane; access time, 10 μs

Auxiliary Drum Storage:
12 groups each of 2,048 registers on magnetic drum; single word or block transfer to and from MS; average access time to single word or block: 8.5 msec within a group, 16 msec to select new group; block transfer rate, 64 μs/word.

Test Storage:
32 toggle-switch registers; 5 flip-flop register (interchangeable with any 5 toggle-switch registers).

SPEED (in microseconds)

Addition:
To get one number from MS, add it to one in Arithmetic Element (AE)   35
To get two numbers from MS, add them, and transfer answer to MS   100

Multiplication and Roundoff:
To get one number from MS, multiply it by one already in AE   50
To get two numbers from MS, multiply them, transfer product to MS   120

TERMINAL EQUIPMENT

Punched Paper Tape and Typewriters:
Flexowriter 7-hole tape (6 information, 1 index); 6-binary-digit code for letters and decimal numbers. Input to computer: mechanical tape reader (106 ms/line, 318 msec(word), photoelectric tape reader (7 msec/line, 21 msec/word); output: type punch (93 msec(line, 279 msec/word), printers (about 135 msec/character, up to 900 msec for carriage return).

Magnetic Tape:
Parallel-serial storage of binary digits in 3 pairs of nonadjacent channels (2 information pairs, 1 index pair). Redundant recording in pairs minimizes tape-flaw errors. Maximum density, 100 lines (200 binary information digits) per inch; speed, 30 inches per sec. Coded tape recording of computations enables printer to operate independently of computer.

Oscilloscope Display:
Five modified Dumont 5-inch oscilloscopes and several 16-inch magnetically deflected CRTs are available for display. X and Y axes each have 2,048 discrete positions (about 350 μs for point or vector set up and display, about 480 μs for character set up and display). Fairchild camera, automatically controlled by computer, can be used with either type.

Buffer Drum:
Magnetic drum acts as temporary storage of input and output data arriving at computer in random and asynchronous manner from multiple sources and leaving for various output devices.

Table 2.2: Characteristics of Whirlwind as of 1954 (see [MANN et al. 1954][p. 1-2])

Only three years later, in 1954, Whirlwind had been expanded with a plethora of peripheral devices and the error-prone electrostatic memory (more about that later) had been replaced by the newly invented core memory – the laboratory oddity had finally evolved into a stable computer system that could be used for research and was no longer the focus of research itself. Table 2.2 shows the setup of Whirlwind in 1954 when it had reached a rather stable state. The following chapter will now describe Whirlwind and its various subsystems in more detail.

3 Whirlwind

Whirlwind was huge – even compared with other computers of its time – and Whirlwind was expensive, more expensive than most other machines built in the late 1940s and early 1950s. ENIAC, for example, had cost about $ 600,000, *Hurricane* – a Raytheon development – came with a price tag of $ 460,000, the famous *IAS computer* cost $ 650,000 etc. Whirlwind was of another class: The overall project amounted to $ 4,500,000 – more than a small fortune back in the 1950s.[1] From today's perspective this money was well invested given all the direct and indirect spin-offs of Whirlwind and its successor, as will be seen in later sections. Nevertheless, back in Whirlwind's days there was quite some argument that its development was *"out of proportion"*.[2] The immense costs of developing and building Whirlwind were caused mainly by the moving-target type of its requirements (started as an analog aircraft and flight simulator, redesigned as a stored-program digital computer capable of real-time operation and finally used for air defense applications) and the necessity of developing many fundamental technologies like core memory which were pioneered in this machine.

Whirlwind was an impressive sight as SEVERO M. ORNSTEIN remembers:

> *"[Whirlwind] filled several floors of the Barta Building at MIT and great bundles of intestine-like cables traversed holes in the floors and walls between rooms."*[3]

Since it was pretty much a work-in-progress from its very beginning to its decommissioning, the following sections describe not a particular state of this remarkable machine but the basic technologies and design decisions that were employed over time.

3.1 Basic architecture

A rather common feature of Whirlwind from today's perspective is the fact that it was a purely binary machine – back then this was a feature that was vividly disputed as there were many computers in operation and development which used a BCD representation of numbers, biquinary codes and other more or less abandoned and forgotten coding schemes. In contrast to modern computers, Whirlwind used the so-called *one's complement* to represent signed integer values while today the *two's complement* dominates.

---

[1] See [FORRESTER 1988][p. 13].
[2] See [REDMOND et al. 1980][p. 166].
[3] See [ORNSTEIN 2002][p. 8].

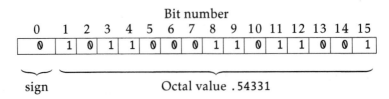

Figure 3.1: Representation of the octal value .54331 in Whirlwind (see [MUHLE 1958][p. 7])

The idea of one's complement is quite simple: To change the sign of an integer value (of 16 bits in the case of Whirlwind) all bits are just inverted, so while 0011011000100111 represents the decimal value +13863, its negated form 1100100111011000 corresponds to -13863 in one's complement. As straightforward as this scheme is, there are two effects which must be taken into account: The so-called *end-around carry* during addition and the fact that there are two different representations of the value zero.

The end-around carry is easy to implement: Whenever a carry is generated from the place of the *Most Significant Bit (MSB)* in an addition (or subtraction) it has to be added back to the *Least Significant Bit (LSB)* of this preliminary result in a second step to form the final result.[4] Let us assume that Whirlwind should compute the result of 17-5:

```
   0000000000010001
-  0000000000000101
   ...
```

Instead of subtracting this last value, its one's complement is added which will generate a carry from the addition of the two leftmost bits, the MSBs of the two values. This leftmost carry bit must then be wrapped around and added to the rightmost bit in a second step:

```
     0000000000010001
+    1111111111111010
   1 0000000000001011
     0000000000001100
```

Since complementing a value, i.e. inverting it in a bitwise fashion, is equivalent to negating it, there are two distinct representations of zero, namely +0 and -0 corresponding to the binary numbers 0000000000000000 and 1111111111111111.[5] The implementation of Whirlwind guaranteed that the result of additions and subtractions which yield zero is represented as -0. The value +0 is only generated when computing (+0)+(+0) or (+0)-(-0).

The structure of a Whirlwind *machine word* is shown in figure 3.1. The basic machine word is 16 bits in length with the bits numbered from 0 to 15 from left to right which is a bit awkward from today's perspective as the weight of a bit position i is determined

---

[4]This is not necessary when two's complement is used instead of one's complement, so in today's arithmetic-logic units there is no need for dealing with an end-around carry.

[5]This effect is also due to the one's complement – in two's complement, there is no -0.

3.1 Basic architecture

Figure 3.2: Sketch of Whirlwind (see [RATHBONE 1951][p. 1])

by 2^{15-i} and not simply by 2^i. Such 16 bit values **x** are treated as normalized numbers[6] with the following property: $-1 + 2^{-15} \leq x \leq 1 - 2^{-15}$. The smallest possible difference between two such values is 2^{-15} as defined by the 16 bit word length.

Figure 3.2 shows a sketch of the physical layout of Whirlwind. The computer itself occupied four large rows of racks while the controls were located in an adjacent control room. Figure 3.3 gives an impression of the control room. As impressive as the plethora of display lights and switches are, they were a mere necessity back then since they were the only means of debugging programs as well as the hardware from a central location. The person on the left is standing in front of the rack containing the marginal checking equipment[7] with an oscilloscope mounted

Figure 3.3: Whirlwind control room (courtesy of the Computer History Museum)

---

[6]Today, machine words are treated as signed or unsigned integers but not as normalized numbers between +1 and −1. The decision to interpret every value as such a normalized value in Whirlwind shows the linkage between the Servomechanism Laboratory with its roots in analog computing and the later DCL and its follow up organizations, since normalized numbers like these are traditionally used in analog computing. [EVERETT et al. 1947][p. 5] points out that the implementation of floating-point numbers was not considered as an alternative to this fixed-point representation since *"the advantages of the floating-point system are outweighed by the extra complexity of the computing equipment and the reduced computing speed."*

[7]See section 3.9.

Figure 3.4: Partial view of two Whirlwind racks (courtesy of the Computer History Museum), persons (from left to right) are JAY W. FORRESTER, NORMAN H. TAYLOR (11/15/1916–02/27/2009), JOHN A. O'BRIEN, CHARLES LEROY CORDERMAN (1924–1979), and NORMAN L. DAGGETT (05/19/1921–08/22/2012)

prominently in the top location. Right next to this rack are a number of flip-flop registers which can be set manually and are mapped into the memory-space of the machine. Some of the many indicators and switches are associated with so-called *computer alarms* which were triggered by error conditions either within the computer, like a timeout when accessing an input/output device, or errors resulting from erroneous programs or data like overflow conditions etc.[8]

Figure 3.4 gives an impression of the layout and structure of Whirlwind's circuitry where two rows of equipment in the machine room can be seen. In contrast to other contemporary machines, Whirlwind was constructed with ease of maintenance and extension as one of the prime objectives in mind:

> "[...] FORRESTER's insistence on reliability led to a clumsy-looking mechanical design. He insisted on a two-dimensional layout to make every component available for replacement without unplugging anything. This yielded a machine that had all the elegance of an elephant. FORRESTER's application of the 'systems approach' produced a monstrous mechanical design but one that had the key attribute FORRESTER desired – easy maintenance."[9]

---

[8] Quite noteworthy is the *continuous alarm* which indicated that various other computer alarms were generated continuously which often was the result of a computer failure (see [MUHLE 1958][p. 102])).

[9] See [JOHNSON 2002][p. 133].

3.1 Basic architecture

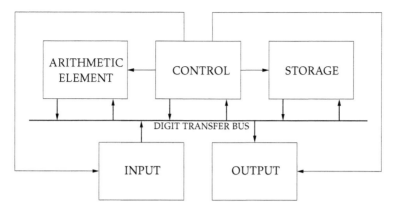

Figure 3.5: Basic structure of Whirlwind (see [EVERETT 1951][p. 71])

The large frame in the middle of the picture holds the electrostatic storage tube system which was later replaced by the world's first production core memory. The storage tubes are mounted in pairs behind the large doors in the middle of this frame. The circuit elements in the top and bottom rows contain the necessary electronics for addressing these memory elements, controlling the storage of data and amplifying and reshaping the signals read from the storage tubes.

Figure 3.5 shows the overall logical structure of Whirlwind.[10] Quite similar to simple modern computer systems, there is a central bus, the *digit transfer bus*, which interconnects the main units of the computer. The overall control of the various subsystems is implemented by means of a microprogram, i.e. every instruction the computer can be programmed to execute is broken down into a sequence of much simpler actions like loading a register, toggling a bank of flip-flops for complementing a number and the like. This sequence of microinstructions is controlled by the *control element*, shown in the upper middle of figure 3.5.

EVERETT notes that

> "[Whirlwind] is basically a simple, straightforward, standard machine of the all-parallel type. Unfortunately, the simple concept often becomes complicated in executions, and this is true here. WW's control has been complicated by the decision to keep it completely flexible; the arithmetic element by the need for high speed, the storage by the use of electrostatic tubes, the terminal equipment by the diversity of input and output media needed."[11]

This observation is especially true for the overall control of the computer. Instead of a single, centralized control element as shown in figure 3.5, a distributed control system was necessary consisting of a central control which also included control of the *arithmetic element*,[12] storage control (this was later simplified by the replacement

---

[10][EVERETT et al. 1947] contains the detailed block diagrams of Whirlwind.
[11]See [EVERETT 1951][pp. 72 f.].
[12]In today's parlance, this would be the *arithmetic logic unit*, ALU for short.

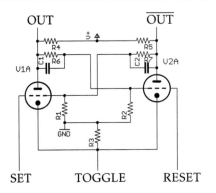
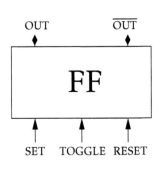

Figure 3.6: Simplified basic flip-flop circuit as used in Whirlwind (see [ADAMS et al. 1951][p. 5])

Figure 3.7: Flip-flop symbol as used in the Whirlwind schematics

of the cumbersome and error-prone electrostatic storage tubes with core memory), and terminal equipment control which took care of the variety of terminal equipment connected to Whirlwind.

The basic circuits of Whirlwind represent the state of the art of the early 1950s but there are some surprises for today's computer circuit designers when closer examining Whirlwind's schematics. Let us have a look at the principle of operations of a basic flip-flop of the type used in Whirlwind. A simplified schematic is shown in figure 3.6: Its three inputs are on the bottom with set and reset on the left and right respectively. The input in the middle, which is connected to both cathodes of the two triodes, is used to toggle the flip-flop.[13] The two outputs, one being the inverse of the other, are connected to the plates of the triodes shown at the top of the schematic. Figure 3.7 shows the generalized symbol for such a set-reset-toggle flip-flop as used in the block diagrams of Whirlwind.[14]

As simple as this basic flip-flop circuit is, there is a catch: The output lines are at a rather high potential even when the corresponding tube is conducting. So either some means of clamping by semiconductor diodes is necessary[15] or AC-coupling by means of capacitors has to be employed. AC-coupling, on which Whirlwind relied, is an easy method to get rid of disturbing DC-levels of a signal by a series-capacitor. This technique is often used in analog electronics but in a digital computer there is a problem: The coupling capacitor must maintain a certain charge for the circuit to operate – therefore it is necessary that every flip-flop coupled capacitively to another circuit must change its state more often than some minimum rate depending on the size of the capacitor etc. The problem with a digital computer is that there is no guarantee that a flip-flop changes its state within a fixed time-interval, so without additional means this simple capacitive coupling scheme would not work. Thus Whirlwind contained special circuitry which stopped the computer every 16 μs, toggled every flip-flop in the machine twice, and then commenced computation again. This process has been known as *restoration*.[16]

---

[13] This circuit is quite identical to the flip-flop stages used in ENIAC as described in section 2.2, figure 2.7.
[14] See [EVERETT et al. 1947].
[15] Whirlwinds successor, AN/FSQ-7, implemented such a clamping scheme.
[16] See [ADAMS et al. 1951][p. 6].

3.1 Basic architecture

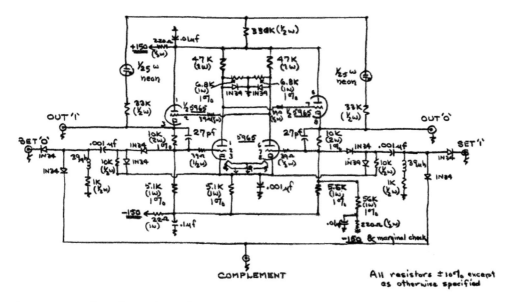

Figure 3.8: High-speed (4 MHz) flip-flop type 5965 (see [BOYD 1953][p. 4])

Figure 3.8 gives an impression of the complexity of a vacuum tube based high-speed flip-flop. The circuit shown was developed in 1953 by HAL BOYD and was designed to drive up to six or seven so-called *gate tubes*[17] per output at a maximum pulse repetition rate of 4 MHz. Four instead of only two triodes are now at the heart of this circuit.[18]

The set and reset inputs are decoupled with 1N34 diodes, which allows the complement input to drive both inputs simultaneously without side-effects to the circuitry driving the two other inputs. The two triodes at the top are cathode followers, pulling the outputs connected to their cathode to plate potential when the control grid is non-negative. These cathode follower stages in turn control the grids of the two lower triodes which form two inverters driving the cathode followers in a crosswise fashion thus completing the flip-flop circuit. The two neon lamps are a debugging and maintenance feature and show the current state of the flip-flop.

3.1.1 Arithmetic element

The *arithmetic element* is the unit in which all actual computations take place; it contains the necessary equipment to store values in a so-called *Accumulator (AC)* and features two additional 16 bit registers, namely A register (AR) and B register (BR). Adding two binary numbers is fairly simple when no high-speed operation is required.

---

[17] These gate tubes were special pentodes 7AK7. A pentode has five electrodes, compared with the three terminals of a triode. In addition to the control grid, a pentode features two additional grids, the *screen* and the *suppressor* grid. In digital applications pentodes were sometimes used as AND-gates by using two of the grids, namely the control and screen grid, to control the flow of electrons.

[18] The type 5965 is a twin triode for "'on-off' control applications involving long periods of operation under *cutoff conditions*" (see [RCA 5965]), a tube explicitly targeted for digital computers.

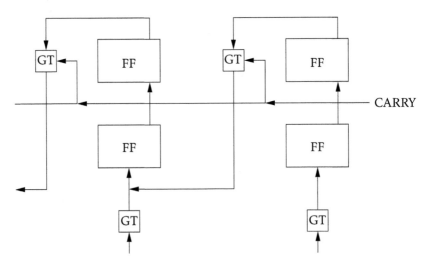

Figure 3.9: Low speed binary adder (see [EVERETT et al. 1947][fig. 14])

Figure 3.9 shows the block diagram of two-stages of a simple low-speed binary adder: The lower flip-flops have their complement-input connected to gate circuits, so they act as simple 1 bit counters, effectively implementing a modulo-2-operation. When the flip-flops are all reset, the first number can be loaded by controlling the complement-inputs bit by bit with the number to be loaded. Adding a second number works the same way – every flip-flop which receives a pulse at its complement-input will toggle its state. If such a flip-flop was reset before, it will now be set, if it was set, it will be reset, and it will generate a pulse at its upper output in this case which represents the carry bit of this particular bit position. These carries are then stored in the upper row of flip-flops. These carry bits are then propagated by a pulse on the line denoted by CARRY which enables the gate tubes GT to the corresponding adder stages to the left.

This simple circuit is, of course, way too slow for a high-speed computer like Whirlwind. A worst-case addition like 11...111+1 will generate the maximum number of carries, namely 16, and each carry propagation step requires one pulse on the CARRY line. So in effect, repetitive CARRY pulses must be generated until all of the carry flip-flops in the upper row are in their reset state. Although statistically there are only about four carries to be considered when adding two 16 bit values, this is of not much help since such an implementation would yield additions of varying duration which would prevent real-time operation of the computer.

Accordingly, another scheme was devised to be used in Whirlwind as shown in Figure 3.10.[19] The idea is to propagate carry information from right to left as quickly as possible through as many stages as possible. The main difference between the circuits of figure 3.9 and 3.10 are the *high-speed carry gate tubes* which receive carry information and the state of the partial sum flip-flop. If this partial sum flip-flop (bottom row) already contains a 1 any carry information coming in from the right can be propagated to the next bit on the left since a carry would be generated from this adder stage in any

---

[19]See [EVERETT et al. 1947][p. 26].

3.1 Basic architecture

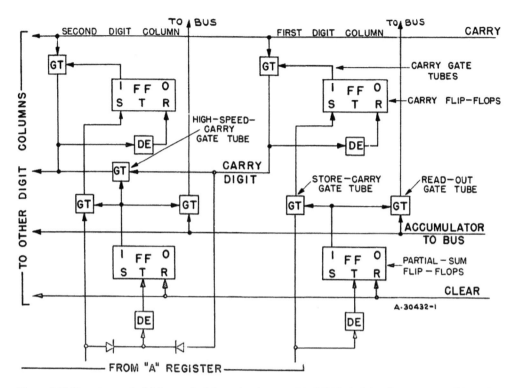

Figure 3.10: Two stages of a high-speed adder as implemented in Whirlwind (see [EVERETT 1951][fig. 31])

case. When the carry signal propagated by the carry gate tubes reaches a stage containing a zero in the corresponding accumulator bit, it will be suppressed since adding a one to a zero does not generate a new carry.[20]

Adding two binary values using this high-speed adding circuit now requires only two steps, assuming that the accumulator, i. e. the lower row of flip-flops, has already been loaded with the first value by means of the complement-inputs of its flip-flops: First, the second value is added in a bitwise fashion. Unlike the carry generation scheme of figure 3.9 where the overflow output of the partial sum flip-flops was used to generate the carry bit, this information is generated in this high-speed adder by means of a *store-carry gate tube* which receives the last output of the corresponding accumulator bit and the bit to be added. To prevent the accumulator bit from changing during the generation of the carry set pulse, the input to the accumulator flip-flop is delayed by a *delay element*, marked with DE in the schematic.

After this step, the accumulator flip-flops in the lower row contain the partial sum while the carry flip-flops contain the corresponding carry bits. The second step now only requires a single pulse on the CARRY line shown on top in figure 3.10 instead of a series of such pulses. This will cause two things: If there is a carry set for a particular bit position within the adder, it will be added to the next bit on the left. In addition

---

[20]In fact, this scheme implements what is known as *carry look-ahead* today.

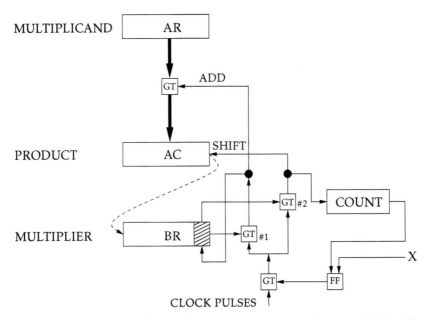

Figure 3.11: Structure or Whirlwind's Arithmetic Element (see [EVERETT 1951][p. 73])

to this a set carry is propagated as far to the left as possible, i. e. until an accumulator flip-flop is reached which is in the reset state.

Figure 3.11 shows the overall structure of the arithmetic element. Its main elements are the AR, which consists only of 16 flip-flops without any additional logic; the accumulator AC, which contains the high-speed adder described above and performs addition and shift operations; and the BR which is used for multiplication and division. All data coming from the digit transfer bus enter the arithmetic element via the AR. In case of subtraction, the complement of the subtrahend is formed in the AR by selecting the negated flip-flop outputs for loading the AC. The BR can be loaded only via the AC and the only way to read out its contents is by performing shift operations. Therefore the BR and AC are treated as a single 32 bit register by some instructions like shifts.[21]

The instruction ca, short for clear and add, is now executed roughly like this: First, the AR and AC are cleared. The AR is then loaded from the digit transfer bus. In the third and last step its contents are added to the cleared AC effectively loading that particular value into AC. Implementing the ad instruction (short for add) is quite similar with the difference that the AC will not be cleared in the first step.

Multiplication is a bit more complicated.[22] The basic idea is to first load the multiplier into BR via AR and AC. Then the multiplicand is loaded into AR. Since multiplying numbers in one's complement representation is highly complex when negative values are involved, the multiplication implemented only works for positive values. Using

---

[21] See [EVERETT et al. 1947][pp. 21 ff.] and [MUHLE 1958][p. 9].
[22] See [EVERETT et al. 1947][pp. 29 f.].

3.1 Basic architecture

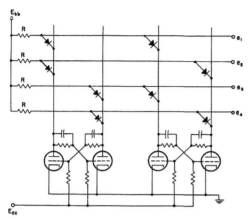 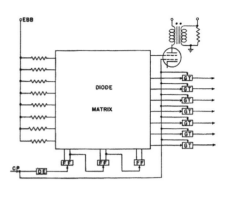

Figure 3.12: Binary decoder, called *switch* in Whirlwind's parlance (see [EVERETT et al. 1947][fig. 9])

Figure 3.13: Time pulse distributor (see [EVERETT et al. 1947][fig. 10])

the complementing ability of the AR it is made sure by conditional control of the complement operation that only positive values are loaded as multiplier and multiplicator. After finishing the multiplication of these absolute values, a sign correction must be performed based on the signs of the original values. The counter shown on the right in figure 3.11 takes control of the multiplication itself, which is implemented as a series of 15 add and shift operations controlled by the rightmost bit of BR which is shifted during each such step. The counter shown on the right controls the shifting process and stops the multiplication after 15 cycles.

3.1.2 Control

As already mentioned, most of Whirlwind's internal operation is controlled by a microprogrammed control unit which will be described in the following. Basically, every machine instruction can be broken down into a sequence of control signals which cause actions like loading the AR from the digit transfer bus, loading the AC from the AR and so on. At the heart of such a control unit are binary decoders, counters and some form of read-only memory – most of which were implemented as diode matrices in Whirlwind.

Figure 3.12 shows the principle of operation of a binary decoder, called *switch* back in Whirlwind's days. At the bottom are two toggle flip-flops, so this circuit is a 2 bit decoder with four output lines since the two flip-flops can be in four distinct states at any time. Let us assume that the right triode of each of the flip-flops is active, i.e. drawing current through the diode matrix and its associated anode resistor. In this case current flows through the resistors of the output lines e_2, e_3 and e_4. Thus these lines will be at a much lower potential as line e_1. So this state of the flip-flops, corresponding to the binary value 00, activates output line e_1.

Accordingly, setting the flip-flops to 01, which causes the right tube of the left and the left tube of the right flip-flop to conduct, draws current through the resistors associated with the output lines e_1, e_3 and e_4 by means of the diode decoder matrix. So

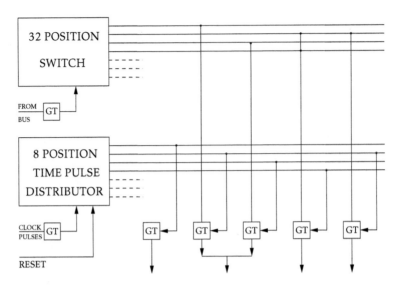

Figure 3.14: Basic control circuit of Whirlwind (see [EVERETT 1951][p. 72] and [MANN et al. 1954][pp. 2-10 f.])

in effect, line e_2 is active etc. Combining such a decoder matrix with a chain of flip-flops acting as a binary counter yields the *time pulse distributor* shown in figure 3.13. The counter is clocked by pulses on the line denoted CP. Apart from incrementing the counter by one, these clock pulses are used to activate the gate tubes on the right of the schematic. So for each counter position a pulse is generated on only one of the output lines and only for the duration of the clock pulse.

Figure 3.14 shows the principle of operation of Whirlwind's central control circuit. On the left are two diode decoder matrices as described above. The one labeled *32 position switch* is fed with the 5 bit *opcode*[23] of the current instruction word, while the *8 position time pulse distributor* has the structure shown in figure 3.13. The pulses controlling Whirlwind's circuitry are derived from the outputs of these two decoder matrices in conjunction with a number of gate tubes:

> "There are 120 gate tubes on the output of the operation control. Pulses on the 120 output lines go to the gate drivers, pulse drivers, and control flip-flops all over the machine; 120 is a generous number. The suppressors[24] of these gate tubes are connected to vertical wires that cross the 32 output lines from the operation switch. Crystals are inserted at the desired junctions to turn on those gate tubes that are to be used for any operation."[25]

These gate tubes combine the outputs of the opcode decoder and the time pulse distributor to generate the necessary sequences which in fact implement the various in-

---

[23] An opcode is the bit pattern which determines the instruction to be executed by a computer. Whirlwind opcodes had a length of 5 bits, so $2^5 = 32$ different instructions could be specified.

[24] The second grid used to control the flow of electrons in the tube.

[25] See [EVERETT 1951][p. 73].

structions Whirlwind can execute. By using such a scheme it is possible to generate more than one control pulse at one time, so effectively, Whirlwind might be classified as having a *horizontal* microcode. Figure 3.15 shows an excerpt of this microcode, the sequence of actions necessary to perform an add operation. As can be seen, the control lines are distinguished into two groups named *program timing* and *operation timing*. The program timing group controls the program counter, the storage from which instructions are read etc., while the operation timing group controls the arithmetic element, the connections to the digit transfer bus, the various registers and so on.

Obviously, the program timing signal group runs through the same sequence of pulses for every instruction being executed, so the operation timing group solely determines what actions will be performed for an instruction. It is interesting to note that the instruction fetch and execution overlap each other. If both activity groups were arranged in a strictly sequential fashion, eleven time pulses would be necessary to accomplish the necessary operations for an add instruction. Since the time pulse distributor can only generate eight time pulses per instruction, there must be some overlap which is possible since the preparations for executing a new instruction (such as addressing the storage system, reading data from memory etc.) have to be completed before any actual computation may take place.

The microcode shown in figure 3.15 is quite instructive regarding this concept of overlapped execution: The first pulse of the operation timing group is generated in clock cycle number seven, when all necessary actions of the program timing group have been executed. Since the control of the arithmetic element for computing a sum requires four steps, the pulse distributor would have to be capable of generating more than only eight successive time pulses if there was no overlap. With overlap, the generation of the operation timing group pulses continues well into the next cycle during which the program timing group already fetches the next instruction from memory.

Figure 3.16 gives an impression of the central control matrix of Whirlwind. Clearly visible are the row lines driven by the decoder and the column lines which drive the gate tubes which in turn generate various distinct control signals which spread out all over Whirlwind.

3.2 Storage

An abundant amount of memory is taken for granted in today's computers, but back in the 1950s, main memory caused much headaches for the hardware developers. Using flip-flops for main memory was not an option due to the immense amount of hardware that would have been necessary even for modest amounts of memory like 256 words of 16 bits each. Some early computers used delay line memories often based on a medium like a long, mercury filled tube to store a series of acoustic pulses (compression waves). These pulses were sent into such a tube by means of an ultrasonic transducer and picked up at the other end after some fixed time-delay determined by the medium and the length of the tube end by a microphone. These signals were then amplified and reshaped and sent into the delay line again, so all one had to do to address a certain bit, was to wait until the bit just came out of the receiving end of the

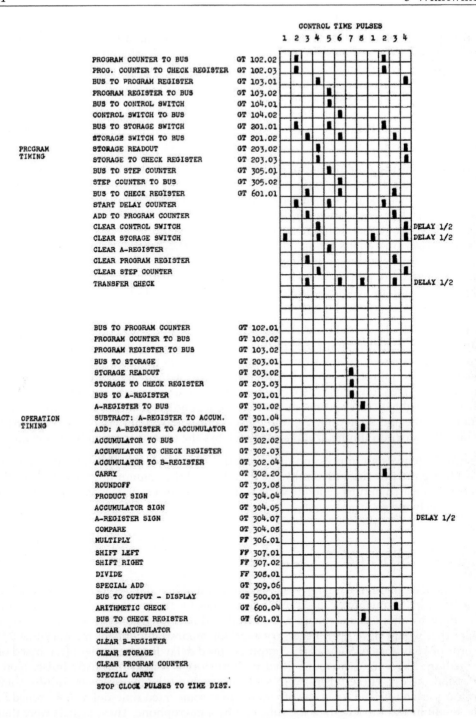

Figure 3.15: Sequence of microinstructions to perform an add instruction (see [EVERETT et al. 1947][fig. 72])

3.2 Storage

Figure 3.16: 32 position diode switch and control matrix of Whirlwind as it was installed in 1949 (see [FORRESTER 1951][p. 46])

delay tube. Variations of this basic idea used magnetostrictive effects in special wires instead of sound waves in mercury, and even magnetic drums like those used in the venerable IBM 650 computer can be seen as some form of a delay line memory with some justification. All of these implementations have the main drawback that they do not really allow for *random* memory access. Depending on the current state of the delay line memory the access time for a bit stored at a certain address varies widely, making real-time processing impossible.

So it was clear from the very beginning that Whirlwind needed another form of storage for its main memory system. Eventually two different technologies were developed and tested: At first, electrostatic storage tubes were developed which proved to be far too unreliable and requiring maintenance far too often to be useful for Whirlwind's ambitious goals. Second, so-called *magnetic core memory* was developed which turned out to be the silver bullet of memory systems until the development of semiconductor memory chips in the 1970s.

3.2.1 The MIT Storage Tube

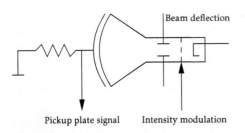

Figure 3.17: Basic setup of a WILLIAMS-tube memory (see [WILLIAMS 1962][p. 12-34])

The idea of using a cathode ray tube for storing digital data has been pioneered by FREDERIC CALLAND WILLIAMS[26] and TOM KILBURN[27] in the 1940s. Figure 3.17 shows the simplified setup of a so-called WILLIAMS- or WILLIAMS-KILBURN-tube: At its heart is a more or less conventional cathode ray display tube shown with its cathode on the right, followed by a control grid to modulate the electron-beam's intensity and a pair of deflection plates.[28] Mounted in front of the face plate of the tube is a so-called *pickup plate* which is connected to ground by a resistor.

When the electron-beam is directed to a certain point on the tube's face plate, it will cause the emission of secondary electrons. In fact, the secondary emission ratio is considerably greater than unity, thus the point to which the electron-beam has been directed by the deflection plates will become positively charged since the secondary electrons are collected by the acceleration anode (not shown in figure 3.17) of the cathode ray tube. The creation of such a positively charged spot is called *digging a well*.

If the electron is directed in the vicinity of an existing well it will create (dig) a new well while the secondary emission electrons from this operation tend to fill the well next to the current location. Since digging a well is a much faster operation than filling another nearby well, there is a shift of charges on the glass front plate of the tube. This charge shift in turn generates a rather tiny signal on the pickup plate which can be

---

[26] 06/26/1911-08/11/1977
[27] 08/11/1921-01/17/2001
[28] Of course, a real WILLIAMS-KILBURN-tube requires two deflection plates to address a two-dimensional area.

3.2 Storage

amplified. So with a precise deflection of the electron beam, charge patterns can be written onto the screen's surface by digging or filling wells representing individual bits.[29]

Although such electrostatic storage tubes were used in a variety of early computers – even some production machines like the IBM 701 – it was deemed unsuitable as main memory for Whirlwind. FORRESTER remembers that it

> "inherently lacked the high signal levels, the high signal-to-noise ratio, the ability to give good signals from the noise, that we would require for our high-reliability application."[30]

Accordingly, development of a novel electrostatic storage tube began in 1947.[31] Figure 3.18 shows the final basic structure of these so-called MIT-storage tubes which were developed and built in a *tube shop* set up specifically for this task by the DCL. In contrast to the basic WILLIAMS-KILBURN-tube this storage tube featured two electron guns: One *high-velocity gun* which emits a finely focused beam of rather high-energy electrons and is used to read and write individual bits, and a *holding gun* which generates a wide low-energy electron beam covering all of the storage surface. This *holding* beam is operating continuously and replaces charges lost due to unavoidable leakage. In addition to this, the charge patterns are not stored on the glass front plate of a typical display tube but instead a *mica* dielectric, holding a beryllium mosaic is used. This intricate face plate design has much superior properties than a simple glass plate. The signal pickup plate is a thin silver film on the back of the mica sheet.

Beryllium was used to form the storage mosaic because of its high ratio of secondary electron emission. The mosaic consisted of individual islands made of Beryllium which were well insulated against each other due to the mica substrate. Individual bits were stored as charge patterns, where a 1 was represented by a charge at the potential of the *collector screen*, i. e. at +100 V, while a 0 corresponded to a potential of the holding gun's cathode (0 V).

To write a 1, the high velocity electron beam is deflected to the desired spot on the Beryllium mosaic by means of two pairs of deflection plates.[32] There it will cause secondary emission of electrons due to the properties of the Beryllium mosaic, effectively charging the selected spot to the potential of the collector screen which absorbes these secondary electrons. Writing a 0 works quite similarly with the difference that the signal plate's potential is temporarily elevated to +100 V during the write operation. When the write operation is completed, the pickup plate is lowered again to ground potential, effectively shifting all charge patterns on the storage mosaic accordingly.

---

[29] In fact, there were many different schemes in use in the 1940s and 1950s for operating such storage systems, called *dot-dash, double-dot, focus-defocus* etc. See [BELL et al. 1949], [WILLIAMS 1962] and [STIFLER et al. 1950][p. 364 ff.] for more in-depth information about this. [ECKERT 1951] contains a thorough description of a practical electrostatic memory system.
[30] See [REDMOND et al. 1980][p. 181].
[31] See [REDMOND et al. 1980][p. 181] and [ECKERT 1997][p. 190].
[32] According to [DODD et al. 1950][p. 4] even this focused electron beam covered 10 to 25 mosaic elements.

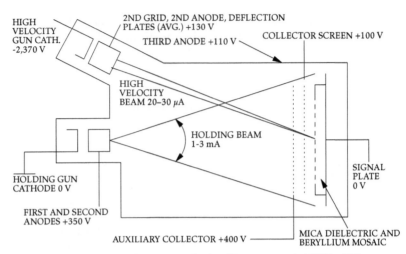

Figure 3.18: MIT 400-series storage tube (see [REDMOND et al. 1980][p. 51])

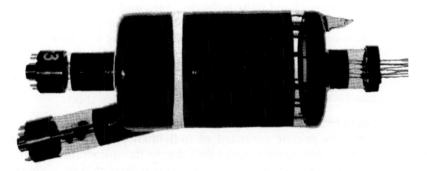

Figure 3.19: MIT Storage Tube (see [DODD et al. 1950][p. 16])

Reading is accomplished in Whirlwind by raising the pickup plate to +50 V so that each change in charge by the high velocity electron beam will result in a signal capacitively induced on the pickup plate. This signal's polarity then contains the information read from the storage tube.

Figure 3.19 shows such a MIT storage tube with its distinctive three necks: The one in the middle left contains the holding gun while the one on the lower left holds the high velocity gun respectively. The neck on the right contains the connections to the signal plate, the collector screen, the auxiliary collector and the third anode. On the upper right the pump port for evacuating the tube can be seen.

As ingenious and promising these tubes were, they were hard to tame and a continuous source of trouble. First of all there were cost and reliability aspects of the overall tube assembly. FORRESTER remembers that each storage tube eventually stored 1,024 bits in a 32 × 32 rectangular charge pattern and cost about $ 1,000 each, having a lifetime of about one month. This resulted in an overall cost of about $ 1 per bit per month.[33] Quite a figure, especially back then. Although reliability of the electrostatic memory system increased from less than a single hour of error-free operation to a couple of hours between 1950 and 1951, this was by far not enough for a computer like Whirlwind.[34] Other persons involved in Whirlwind's development remembered quite vividly the temperamental nature of these tubes:[35]

> "The mean-time-to-failure of these tubes was very short; much too short for a practical system. In fact, in order to get anything near satisfactory performance, the engineers who were putting the memory together had to know the idiosyncrasies of each of the tubes that was used. One of those engineers, ALAN J. ROBERTS, remembers one of the times he was asked to fix a problem with this memory system. One of the 32 tubes in the system had a mechanical fault in the signal plate assembly that was used for writing ones and zeros. Apparently, there was some play in the structure that held the signal plate screen in place. Several times when he was called in, AL went directly to that particular tube, tapped it in a certain way, and cleared up the trouble. The necessity of an engineer's knowing the weaknesses of the individual tubes, practically calling the tubes by name, promised to be the rule, and not the exception, in CRT memory."[36]

After some years of development it became clear that these electrostatic storage tubes would never mature into a usable and useful memory system for a high-speed, high-reliability computer. The *Project Summary Report* no. 35 from 1953 states that

> "[s]ince permanent cures for these maintenance troubles seemed impossible to achieve and since Magnetic-Core Storage was proving itself extremely promising in the Memory Test Computer, it was decided to replace Electrostatic Storage with Magnetic-Core Storage."[37]

Thus the development of storage tubes came to an end in the early 1950s and attention shifted to another, highly-successful memory system based on tiny rings made of ferromagnetic material exhibiting a very peculiar hysteresis curve. This memory technology which literally shaped the whole computer industry during several decades, beginning in 1950, will be described in the following section. Meanwhile, the tube shop shifted its attention to the development of cathode ray display tubes and made contributions to the so-called *Charactron* and *Typotron* tubes which would eventually be used extensively in the display consoles of SAGE.[38]

---

[33] See [EVANS 1983/2][p. 401].
[34] See [JOHNSON 2002][p. 140].
[35] The extreme unreliability of electrostatic storage tubes in general was also a main problem for other machines like the IBM 701, see [HADDAD 1983][pp. 122 f.] although or maybe because these relied on the simpler WILLIAMS-KILBURN-tube.
[36] See [JACOBS 1986][p. 13].
[37] See [PWSR 1953][p. 32].
[38] See [JACOBS 1986][p. 14].

3.2.2 Magnetic core storage

Already back in 1947 other ideas for reliable and fast storage and retrieval of digital data had been explored in parallel to the development of the electrostatic storage tubes. Generally a binary storage element needs to exhibit a highly nonlinear behavior to discriminate reliably between two stable states representing 0 and 1. One of these early attempts conducted by FORRESTER was based on a three-dimensional arrangement of gas-discharge elements similar to neon lamps. It quickly turned out that the nature of the discharges was too erratic for any reliable storage application, so this idea was dropped after only some preliminary studies.[39]

Then, by mere chance, FORRESTER found a promising material with a highly nonlinear behavior. Elements made from this material could even be arranged in two- and three-dimensional setups. FORRESTER remembers:

> "I remember very clearly how it happened. I was reading a technical journal one evening, just leafing through the advertisements in the magazine Electrical Engineering, when I saw an advertisement for a material called Deltamax, which had a very rectangular hysteresis loop. I think it was derived from a material developed in Germany in World War II. [...] When I saw this nonlinear rectangular magnetic hysteresis loop, I asked, 'Can we use it as a computer memory? Is there some way to fit it into a three-dimensional array for information storage?' The idea immediately began to dominate my thinking, and for the next two evenings I went out after dinner and walked the streets in the dark thinking about it[...]"[40]

The idea FORRESTER had is based on a simple transformer like that shown in figure 3.20 which consists of a toroidal core carrying two windings. The left winding may be considered the *primary* and the right the *secondary*. If the core does not exhibit any hysteresis, this setup just behaves like any other transformer: A varying current I_m in the primary winding results in a changing magnetic field which in turn induces a voltage in the secondary winding. The induced voltage mainly depends on the ratio of winding of the primary and secondary.

With hysteresis, things become more interesting. Figure 3.21 shows a typical rectangular hysteresis loop as exhibited by Deltamax and many later materials which were developed explicitly for storage and switching purposes. The x-axis of the hysteresis loop figure shows the current I flowing through the primary winding which yields a magnetic field strength of $H(I)$. The y-axis shows the magnetic flux density Φ of the transformer's core. As can be seen, there is no simple linear relationship between H and Φ, in fact, there are two stable flux densities marked with *one* and *zero* respectively.

---

[39]See [EVANS 1983/1][pp. 297 f.] and a page from JAY WRIGHT FORRESTER's 1949 notebook, see http://www.computerhistory.org/revolution/memory-storage/8/253/982, retrieved 11/28/2013.
[40]See [EVANS 1983/1][pp. 298 f.].

3.2 Storage

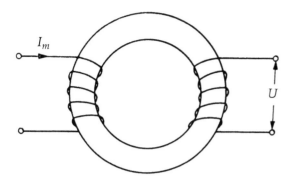

Figure 3.20: Single magnetic core with input and output winding (see [Valvo 1965][p. 11])

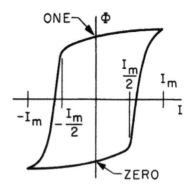

Figure 3.21: Hysteresis of a ferrite toroid (see [Brown 1953][p. 6])

These two states of the core are used to represent one bit of information and will be denoted by Φ_1 and Φ_0 in the following.[41]

Suppose that the core currently has a remanent magnetization Φ_1. To change this state to Φ_0 a current greater than $-\frac{I_m}{2}$ must flow through the primary winding. This will bring Φ even below the value Φ_0. When the current becomes 0 again, the remanent magnetization will reach Φ_0 and stay there. To change this back to Φ_1 again, a current in excess of $+\frac{I_m}{2}$ must flow through the primary winding. Currents I with $I < |\frac{I_m}{2}|$ do not change the current state of the core.[42] A critical parameter for such a magnetic core is the degree of rectangularity R of its hysteresis loop which is defined as

$$R = \frac{\Phi(-\frac{I_m}{2})}{\Phi(H_m)}.$$

Now that a bit can be stored as a magnetization in a toroidal core the question remains how such a bit can be retrieved later. This is where the secondary winding shown in figure 3.20 comes into play. As in every transformer, a voltage is induced in the secondary by a change in magnetic flux in the core. Due to the rectangular hysteresis curve there are only two stable magnetizations Φ_1 and Φ_0, so only a full remagnetization from Φ_1 to Φ_0 or vice versa results in a significant change in magnetic flux which can then be detected as a voltage across the secondary winding. To determine if Φ_1 had been stored in the core, it is only necessary to try to change its magnetization to Φ_0. If this causes a change in magnetization it is clear that the previous state was set to Φ_1. The only drawback of this scheme is that readout of such a core memory cell is

---

[41] Developing materials showing such a rectangular hysteresis loop was no simple task and many early core samples were unusable since there were no stable states at all or sometimes three states. Just as a side note: Back in 1952 it was proposed to build a ternary computer based on such cores as switching elements, see [Grosch 1952].

[42] In fact, they do change the state but not considerably. A core that was in Φ_1 for example, will go into a state Φ_1^* which is represented by a slightly smaller remanent magnetization. Fortunately, there is a lower bound for Φ_1^* and Φ_0^*, so that further attempts to change the magnetization with insufficient current won't gradually degrade the magnetization of the core beyond these values.

destructive.[43] Accordingly, bits read from such a memory system must be restored in a following step.

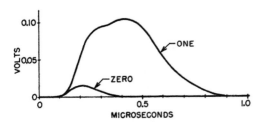

Figure 3.22: Readout from a selected ferrite toroid (see [BROWN 1953][p. 5])

Figure 3.22 shows typical signal curves of the voltage induced in the secondary winding which result from reversing the magnetization of a core, i. e. reading the state Φ_1 by overwriting it with Φ_0, or from not reversing the magnetization when the state already was Φ_0.

Of course, the development of these basic techniques took some time, since not only a variety of proposed core materials had to be tested, but a physical explanation of the effects within the core material had to be developed to gain the necessary insight to build a reliable storage system. One question which received considerable attention was that of the effect currents smaller than $|I_m/2|$ had on the remanent magnetization of the core which was described above.

Figure 3.23: Two-by-two array of magnetic core memory for early testing

Figure 3.23 shows an early core memory setup of FORRESTER's group. Only four cores, which can be seen in the middle of the picture, are driven by suitable line drivers located on the panels below the core assembly. The panels above the cores contain the amplifier stages to process the signal induced in the secondary winding. Depending on the amplitude of the value, the bit just read by magnetizing the core is determined.

Retrospectively, it turns out that the time for inventing a memory system based on magnetic materials with a rectangular hysteresis loop had come in the late 1940s, and it would be difficult to proclaim a single inventor of magnetic core memory. In 1949, JAN ALEKSANDER RAJCHMAN[44] proposed an early version of core memory (see [RAJCHMAN 1957]). Also in 1949, FORRESTER wrote a detailed proposal of a core memory system.[45] Another inventor, FREDERICK W. VIEHE,[46] filed a patent for a magnetic core memory which he had developed in his *"home laboratory"* in 1947.[47] One of the main objectives of this early patent application was stated as follows:

---

[43] Although there was some research on nondestructive sensing of magnetic cores (see [BUCK et al. 1953]), most of the commercially available core memories were based on destructive readout.

[44] 08/10/1911–04/01/1989, a short biography can be found in [HITTINGER 1989].

[45] See [FORRESTER 1950].

[46] 1911–08/14/1960. According to [Redlands 1960][p. 7] he *"left an estate of $ 623.009. [...] VIEHE perished Aug. 14 in the blazing heat of the Mojave desert during a rock hunting trip. [...] An attorney [...] said VIEHE made his fortune through sale of a secret invention, but was sworn to secrecy regarding the invention and sale."*

[47] See [REILLY 2003][p. 162] and [VIEHE 1961].

3.2 Storage

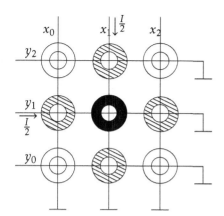

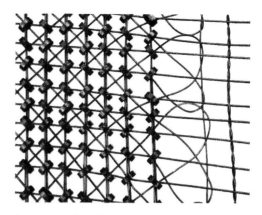

Figure 3.24: Selection of a magnetic core in a two-dimensional matrix

Figure 3.25: Detail photograph of Whirlwind core memory plane (© MARK RICHARDS, courtesy of the Computer History Museum)

> "A further object of my invention is to provide a transformer utilizing a core composed of a material having appreciable magnetic retentivity that may be employed for operating or restoring a trigger circuit in which the transformer is arranged, according to the prior history of the core."[48]

Despite all these forerunners and contemporary inventors, the core memory developed for Whirlwind under the auspices of FORRESTER should become the first practical core memory system. The fact that recurring magnetization attempts with small currents have no severe degradation effect on the remanent magnetization of a core allowed the development of two-dimensional arrangements of cores in form of core *matrices* or *planes*.

Figure 3.24 shows a two-dimensional arrangement of magnetizable cores. Instead of a single primary winding there are two wires crossing each core. To select a particular core for magnetization to Φ_1 or Φ_0, a pair of row- and column-wires will be selected. These two wires will then be connected to current sources delivering currents of $I_m/2$. Only the core at the intersection of these two wires experiences a magnetic field strong enough to change its remanent magnetization, while the other cores on the selected row- and column-wire are only affected by $\frac{I_m}{2}$ and will not change their current state. Accordingly, such an arrangement is called *coincident current core memory*. Such a two-dimensional arrangement of magnetic cores drastically reduces the amount of necessary driver circuits since only $2n$ such drivers are necessary for a matrix holding n^2 cores. Typical sizes of early core memories were $n = 32$ and $n = 64$, so a single core plane held $1,024$ or $4,096$ cores respectively.[49]

What about the secondary winding necessary to read out the current state of a selected core by trying to change its magnetization? This is called the *sense winding* or *sense*

---

[48] See [VIEHE 1961].

[49] As early as in 1956 there was some work toward the development of magnetic thin-film memory systems which would not require threading wires through individual cores, see [CHILDRESS 1956].

wire and is implemented as a third wire threaded through every core. In the simplest case, this sense wire is threaded diagonally through the core plane, but to minimize problems due to crosstalk between the row- and column drive lines, other (often quite intricate) threading schemes have been employed.[50]

Figure 3.25 is a closeup photograph of a single core memory plane from Whirlwind. The cores are located at the intersections of the row- and column-wires connected to the driver circuits. The sense wire with its distinctive diagonal threading scheme is also clearly visible. A closer look reveals that there is an additional, fourth wire running vertically through every core which has not been covered yet.

This is the so-called *inhibit winding* or *inhibit wire*. There is only one such wire per core matrix running up and down repeatedly thus reaching every core. The purpose of the inhibit wire is to block write operations to a matrix which is necessary when a number of such planes are stacked to form a three-dimensional *memory stack*. The basic idea of forming a stack of core planes is to address all planes in parallel and use one plane for one bit of a word. So in case of a 16 bit machine, a stack of 16 planes, each containing 32×32 and later 64×64 cores, would form a memory system capable of storing $1,024$ or $4,096$ words of 16 bits each.

Figure 3.26 shows the basic arrangement of planes in a core stack. To minimize the amount of driving circuitry for the row- and column-wires, all corresponding row- and column-wires of the individual planes of a stack are connected in series, so two currents flowing through a selected pair of wires will affect all cores at the intersection point in each plane. Thus without additional means of control it would not be possible to treat corresponding cores in different planes differently.

Using the inhibit wire, it is now possible to send a current of plus or minus $\frac{I_m}{2}$ through all cores of a single plane. This current will not change the state of any core in that matrix but it will counteract the sum of the currents at the intersection of the selected row- and column-wire, effectively suppressing the write-access to this particular plane. Assuming that all cores in the planes of a stack corresponding to a certain address have been reset to Φ_1, writing a data word to this address is then performed by setting up appropriate currents for the individual inhibit wires and then writing Φ_0 at the selected address. The cores in the planes which have been blocked by the inhibit current will stay in Φ_1 while the other cores will be switched to Φ_0.[51]

Building these core planes which required threading extremely thin wires through thousands of cores turned out to be a nerve-wrecking task which was most often performed by young women.[52] JACOBS remembers a particularly funny anecdote:

---

[50] See [Valvo 1965][pp. 33 ff.] for some examples.

[51] It should be noted that the direction of current necessary to drive the inhibit lines alternates from plane to plane due to the interconnection scheme used as can be seen in figure 3.26. See [STUART-WILLIAMS 1962] for more information about early magnetic core memory systems, drive circuits etc.

[52] It should be noted that VIEHE continued to work on magnetic memory systems. He proposed using *"magnetic particles suspended in a dielectric medium"*. The idea was to form the magnetic storage elements automatically in an existing wire matrix without the need of manually threading thin wires through tiny cores (see [VIEHE 1968]). Unfortunately this never worked reliably enough for a commercial product.

3.2 Storage

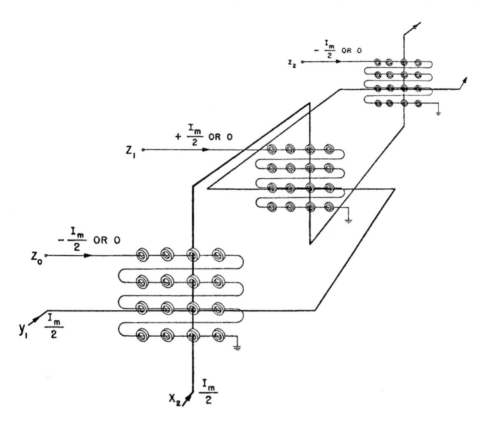

Figure 3.26: Writing to a three-dimensional arrangement of cores (see [MANN et al. 1954][p. 5-2])

> "In the course of core development, visitors came to MIT to see how things were done. JOHN [BANNISTER] GOODENOUGH[53] was giving a contingent from Germany an explanation of the wiring process. John had had some German in his education, and chose to use some German words with this group. In describing the process, he was explaining that young women had been hired to do the sensing wiring. He used the 'liebfrau' to describe the women. He meant to connote young woman, but the word also implies virgin. After the tour, the Germans spent much of their time trying to determine how the liebfraus were selected."[54]

A fully assembled core stack from Whirlwind is shown in figure 3.27. In the middle of the large steel frame resides the actual stack of 17 core planes.[55] The serial connections of corresponding row- and column-wires can be seen on the front of the stack. The two times seventeen coaxial cables connected to the right side of the core stack carry the signals from the sense wires and to the inhibit wires of each plane.

---

[53]*07/25/1922
[54]See [JACOBS 1986][pp. 14 f.].
[55]One core plane was used to hold parity information.

Figure 3.27: First bank of magnetic-core storage in Whirlwind (see [PWSR 1953][p. 33])

Above and below the core stack are the large vacuum tubes at the heart of the row- and column-wire driver circuits. The currents required to switch a core from one stable state to the other were quite large at about 200 mA per line thus requiring power pentodes as drivers. Due to their peculiar shape, the large steel enclosures holding the memory stacks became known as *"shower stalls"*.

Magnetic core memory, the development of which was originally begun mainly due to the extreme unreliability of electrostatic tubes, which turned out to be an engineering disaster,[56] quickly matured into what would become the icon of storage for about three decades. It proved to be unexpectedly reliable and featured short access times:

> *"On August 8 [1953], one bank of [Whirlwind's] electrostatic storage was replaced by a bank of magnetic-core storage. The replacement bank had been operating successfully as part of the [Memory Test Computer (MTC)].[57] The second bank of ES was replaced by a second bank of magnetic-core storage on September 5. The access time of the new storage is 9 microseconds, compared with 25 microseconds for ES. Reliability is also greater, reducing maintenance time."*[58]

---

[56]See [REDMOND et al. 1975][p. 1.06]: *"Even its magnificent, internal, magnetic-core storage had emerged as a desperate, risky, ad hoc engineering solution to the nagging problems of unreliable electrostatic-tube storage."*

[57]*"The original purpose of the construction of the Memory Test computer was, as its name implies, to provide realistic tests of the practicability of the newly-developed coincident-current magnetic code memory. After several months of extensive memory tests on the computer, the original magnetic-memory system was transferred to WWI, abruptly terminating this test program. There existed at this time, however, a growing need for a 'proving ground' for devices and techniques proposed for use in future computers, as well as for additional computing facilities. The activity of the computer has been directed toward filling these needs."*, see [BAGLEY 1954][p. 5]. The MTC was the first computer built by KENNETH HARRY OLSEN, 02/20/1926–02/06/2011, who would later found the *Digital Equipment Corporation*, *DEC* for short. Detailed information about the MTC can be found in [ZIEGLER 1957] and [BAGLEY 1954].

[58]See [PWSR 1953][p. 5].

3.3 Magnetic drums

In its final configuration[59], Whirlwind featured a core memory system of 6,144 memory locations, called *registers*. This core memory had an access time of only 7 µs and consisted of three such shower stalls, two containing 17 core planes of 1,024 cores each and one containing 17 planes of 4,096 cores.

Since early estimates of the amount of memory required to solve the tasks Whirlwind was being built for were conservative, addresses were limited to 11 bits, allowing Whirlwind to address only one out of $2^{11} = 2,048$ memory locations. Accordingly, a scheme had to be implemented which allowed to divide the available 6k words of core memory into so-called *banks* of 1,024 words each. The address space of Whirlwind was also divided into two such banks, called *Group A*, containing addresses 0 to 1,023, and *Group B* for the addresses 1,024 to 2,047. Two of the six memory banks could then be mapped into the Group A and Group B address space[60] using the change fields instruction.[61]

3.3 Magnetic drums

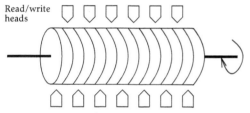

Figure 3.28: Schematic of a magnetic drum

Figure 3.28 shows the basic structure of a so-called *magnetic drum*. At its heart is a rotating drum with a magnetic coating like brown iron oxide,[62] which is driven by a motor. Typical speeds were 3,600 rpm for drums like those used for Whirlwind although other machines, like the IBM 650 drum computer which used a magnetic drum as its main memory system, achieved up to 12,000 rpm. The drum is divided into so-called *tracks* with dedicated read/write heads for each track, thus eliminating any latency for moving clumsy head assemblies.[63] Figure 3.29 shows a typical drum like those used in Whirlwind.

Whirlwind had two different types of magnetic drum subsystems attached: One was called *auxiliary drum*[64] while the other was known as *buffer drum*. The drums for both of these storage systems were constructed and built by *Engineering Research Associates* of St. Paul,[65] *ERA* for short.

The auxiliary drum, which was similar to a standard magnetic drum as that used in the ERA 1101 computer, was used as an external storage system from which programs

---

[59] See [MUHLE 1958][p. 5].
[60] A bank of memory could not be mapped to Group A and Group B simultaneously.
[61] cf for short.
[62] Fe_2O_3
[63] Only a few drum systems had fewer heads than tracks which saved over all cost of the drum assembly but required the introduction of a sled onto which the heads were mounted. This sled could then be positioned over any desired track. This movement required a considerable amount of time and thwarted the main advantage of a magnetic drum.
[64] A detailed description of the auxiliary drum and its programming can be found in [FORGIE 1953].
[65] Minnesota

Figure 3.29: Typical magnetic drum with scratches caused by heads actually touching the magnetic surface due to misalignment (© Mark Richards, courtesy of the Computer History Museum)

and data could be loaded into the main memory of Whirlwind which still was based on the electrostatic storage tubes at the time the drum systems were developed.[66] The rationale of including such a device into Whirlwind was based on the following observation:

> "It has long been apparent that the present capacity or even the projected final capacity of electrostatic storage was inadequate for many of the complex problems confronting the computing field today. Magnetic tapes and drums are storage media which can be used to achieve relatively large storage capacity at moderate cost. Magnetic tape offers essentially unlimited storage capacity, but this advantage is obtained at the cost of a relatively long access time. The magnetic drum offers a relatively large capacity compared to electrostatic storage and an access time much shorter than magnetic tape. Access to a random register on a magnetic drum is generally much slower than access to a register of electrostatic storage, but if a block of registers can be handled in sequence, the average access time per register will be comparable. The capacity of the drum is obviously limited in comparison to magnetic tape, but is still large enough to afford a considerable improvement over electrostatic storage."[67]

One track of this 8.5 inch diameter drum held $2,048$ bits and there were a total of 194 heads installed with a clearance of 0.002 inches between head and drum surface.[68] The

---

[66] Accordingly, all transfers involving the auxiliary drum took place between the main memory and the drum. Transfers could be single word or block transfers (see [Rich 1951][p. 3]).

[67] See [Forgie 1953][p. 3].

[68] This drum, being 18 inches long (see [Stifler et al. 1950][p. 338]), allowed for a total of 208 heads to be installed, but only 194 heads were actually used (see [Forgie 1953][p. 5]).

3.3 Magnetic drums

drum and the enclosure it was mounted in were built from the same aluminum alloy since different temperature coefficients of the drum and the surrounding structure holding the heads would result in a varying clearance which could in turn cause either so-called *head crashes*[69] or at least change the read voltage levels significantly.

These 194 heads were combined into 12 groups of 16 heads each to allow parallel word transfers between Whirlwind and the drum, resulting in a total capacity of 24k bytes for the auxiliary drum.[70] The remaining two heads were used to read timing information which was written to the drum during manufacturing[71] in order to synchronize read and write operations. One of these tracks, the *timing track*, contained 2,048 timing pulses while the other, called *bracket track*, delivered only one pulse per revolution to denote the first row of data on the drum.

The pulses read from the timing track controlled an Angular Position Counter (APC) which was incremented by each pulse and was reset by the pulse read from the bracket track. To address a specific word on the drum to be read or written, 15 bits were written into a storage address register. Eleven of these bits were compared against the current value of the APC while the remaining four bits were used to select one of the twelve groups of 16 heads for reading or writing. Switching the read amplifiers from one group of heads to another was done with a diode matrix resulting in switch times of 30 μs. In contrast to that, the connections between the heads and the write amplifiers were made by a relay matrix which was considerably slower at 24 ms (start of development) and 12 ms in the final implementation.[72] Figure 3.30 shows a block diagram of the auxiliary drum system.

At about 3,600 rpm, the time to read or write one word of 16 bits being directly under the read/write heads was slightly over 8 μs. The maximum access time for a word of an already selected head group was about 17 ms while write access to a word in a different head group took 29 ms, including the time to setup the relay matrix, at most. A block transfer would result in transferring one word every 8 μs from or to the drum which could not be achieved with the electrostatic storage tube memory of Whirlwind. Thus a scheme of interleaving data on the drum was employed: Consecutive words were not placed in consecutive locations on the drum but were spaced eight slots apart, so that every 64 μs consecutive words of a block were transferred. Due to the capacity of 2,048 bits per track, this resulted in the following sequence of addressable bits on consecutive angular positions of the drum: 0, 256, 512, 768, 1,024, 1,280, 1,536, 1,792, 1, 257 etc. This addressing scheme allowed block transfers to run without ever overrunning Whirlwind's memory while reading from the drum, or underrunning the drum while writing, thus significantly speeding up operation in general.[73]

The buffer drum, featuring 186 read/write heads arranged in ten groups of 16 heads for reading/writing data, two heads for reading timing and bracket data, and twelve dual heads for reading and writing meta-data, served a different purpose:

---

[69] A head actually touching the surface of the drum.
[70] See [FORGIE 1953][p. 6] and [PWSR 1951][p. 6-123].
[71] See [FORGIE 1953][p. 12].
[72] See [RICH 1951][pp. 2 f.] and [FORGIE 1953][p. 4].
[73] See [FORGIE 1953][pp. 4 f.].

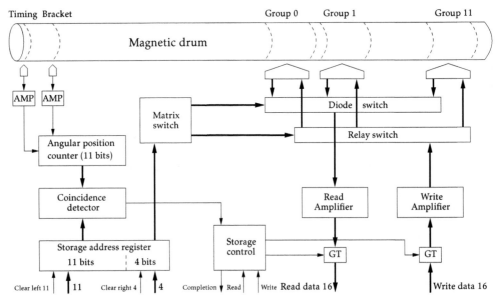

Figure 3.30: Simplified block diagram of the auxiliary drum system (see [RICH 1951][p. 16])

> "The primary reason for existence of the buffer drum is to store asynchronously arriving information until the central computer can absorb it in large blocks. Incoming radar data must be continuously recorded as it arrives, and at the same time the central computer must be able to read the buffer drum essentially at its convenience."[74]

So the buffer drum effectively performed an impedance matching with respect to the extremely different speeds of Whirlwind and some of its input/output equipment. To eliminate the need for the relay switch with its long setup times, the buffer drum featured one write amplifier for each individual write head while two[75] groups of 16 read amplifiers were implemented. Using a diode switch, these read amplifier groups could be connected to each of the ten head groups. Like the auxiliary drum, the buffer drum used an APC driven by signals read from a timing and a bracket track. In contrast to the auxiliary drum, the APC drove the input lines of a diode matrix[76] which implemented a rather flexible interleaving scheme by reordering drum addresses.[77]

Yet, the biggest difference between auxiliary and buffer drum are the additional twelve groups of dual read/write heads of the buffer drum. These were used to determine whether a track contains valid data, called a *filled track*, or whether a track is free and ready to receive data, called *empty track*. This mechanism was necessary since none of

---

[74] See [MUHLE 1958][p. 78].
[75] This may have been expanded to four in a later implementation, the sources are a bit unclear.
[76] See [RICH 1951][pp. 2 f.].
[77] The interleaving scheme of the auxiliary drum was implemented by a mechanism that incremented the storage address register automatically by eight (or nine after one revolution of the drum) whenever the coincidence detector was triggered. This is not shown in figure 3.30, see [FORGIE 1953][p. 15] for more information.

the external drum accesses were under control of Whirlwind and had to be synchronized transparently with its operation.[78]

It is interesting to note that already in 1951 there were thoughts about using the buffer drum to interface displays to Whirlwind, something which should be crucial for AN/FSQ-7:

> "Some consideration has been given to the use of one or more groups on the buffer drum for storage of data to be displayed on a scope. This application is suggested by the fact that the periodic reappearance of a given piece of data under the reading heads on the drum would satisfy the requirement for repeated intensification of an elementary spot in a scope display. In this manner a persistent display pattern could be obtained without having to recalculate all the spots periodically or to rely on cathode ray tubes with long persistence phosphors."[79]

3.4 Magnetic tapes

Whirlwind also featured a simple magnetic tape subsystem, the purpose of which has been described as follows:

> "Magnetic tape will be used for storage of temporary results, storage of subprograms, storage of results to be printed, etc. Equipment consists now of one Raytheon-built magnetic tape assembly, but it is expected that four such units will be available."[80]

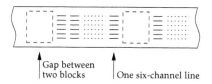

Figure 3.31: Basic structure of a Whirlwind tape

The tape units used were built by Raytheon and used 1/2 inch wide tape on reels holding up to 1,000 feet of tape. Figure 3.32 shows such a tape drive – remarkable features are the take-up and supply reels mounted one over the other. Data was written in *lines* of six bit each on the tape as shown in figure 3.31. Since the tape material available in the early 1950s had been developed for analog recording purposes, blemishes of the magnetic coating were a common phenomenon. In an analog recording these imperfections are not too serious, as long as they do not cover too large of an area. When used as a digital storage medium, these blemishes cause so-called *drop outs* since they cause bits or bit groups to be lost. To counteract this problem, the Whirlwind tape system connected heads 0 and 3, 1 and 4, and 2 and 5 in parallel,[81] so that of the six available channels only three were avail-

---

[78] See [RICH 1951][pp. 3 f.].

[79] See [RICH 1951][pp. 15 f.].

[80] See [HEART 1952][pp. 10 f.].

[81] Studies were performed which showed that these blemishes and dust always caused bits to be lost and sometimes stretched over two adjacent bit positions vertically and horizontally. Thus every third head was connected in parallel instead of paralleling adjacent heads directly (see [PWSR 1951][pp. 6-124 ff.]). It should be noted that using read/write heads connected in parallel was also used many years later in the DECtape tape drives.

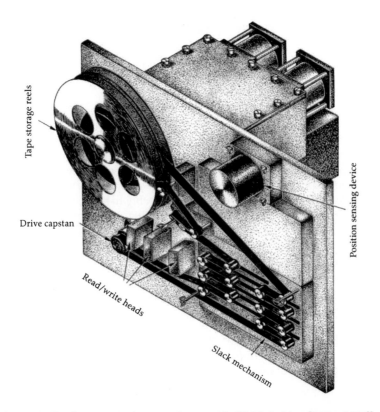

Figure 3.32: Raytheon magnetic tape unit as used in Whirlwind (see [Bloch 1951][p. 54])

able to the input/output interface. Two of these channels were used to store actual data while the third channel acted as an index channel and triggered the flip-flops which stored the data from the two data channels.

The data density of the tapes was 100 lines per inch. Running at a nominal speed of 30 inches per second, this resulted in a typical transfer rate of 375 words of 16 bits each per second. Blocks of consecutive data lines were delimited by gaps 0.2 inches in length, long enough to allow stopping the tape drive mechanism without running into the next block of data. The tape could be run forward or backward while data was written or read back. At any time a tape unit was operating in one out of three modes: *read*, *record*, and *re-record*.[82]

Writing a block of data, i. e. a sequence of lines delimited by gaps, required an instruction sequence like that shown in figure 3.33. The first line of each block of data consists of a special character, the so-called *block mark*, which is written automatically. This mark is used during a read operation to determine the start of a block.[83]

---

[82] In re-record mode a tape which has already been used to store data is overwritten with new data, see [PWSR 1951][p. 19].

[83] See [PWSR 1951][p. 19].

> 1. Select one out of four available tape units, select tape direction, select mode of operation, start selected unit, and write a record block mark (si instruction – section appendix A).
>
> 2. Record one 16 bit word (rc instruction) and wait for completion of the write operation.
>
> 3. Repeat step 2 until all data has been written.
>
> 4. Stop unit (si instruction).

Figure 3.33: Sequence of operations to write a block of data onto magnetic tape (see [HEART 1952][p. 12])

3.5 Paper tape readers, punches and typewriters

Paper tape as a storage medium was already well known and used extensively in telecommunication equipment. It was rarely used for computer applications since most (commercial) computing installations of the 1950s relied on so-called *record* equipment using punched cards. Whirlwind changed this, as it used cheap paper tape input/output equipment instead of mechanically complex punched card machines.

Basically, a paper tape as used in Whirlwind is just a long strip of paper with holes arranged in *lines* of six bits each, quite like those used in the magnetic tape system.[84] A paper tape holds ten lines per inch, one tenth of the data density of Whirlwind's magnetic tape system.

For output two *Flexowriters* were available, each featuring a printer and a paper tape punch.[85] Figure 3.34 shows a typical Flexowriter – its integral paper tape punch is visible on the left.

For input Whirlwind used two paper tape readers: One photoelectric reader built by ERA, and one *Flexowriter* electromechanical reader of older design. While both of these readers could be used in so-called *normal* mode, in which one line of data was read by a single rd instruction, the photoelectric reader also featured an *automatic sexadecimal*[86] mode of operation. A single read instruction issued in this mode of operation caused four lines of data being read automatically. The lower four bits of these lines were then combined into a 16 bit word and transmitted to Whirlwind.[87]

Figure 3.34: Typical Flexowriter (see [Commercial Controls])

---

[84]The paper tape readers and punches actually worked with lines of seven bits but only six were used by Whirlwind's input/output system.
[85]See [BLODGETT 1955].
[86]This would be called *hexadecimal* in today's parlance.
[87]See [HEART 1952][pp. 5 ff.].

3.6   Data transmission

Digital data transmission was a crucial point for the application of Whirlwind to the air defense problem, since it was the only way of connecting remote radar stations to a central computer.[88] Early work on this had been done in the late 1940s by a group lead by JACK VINCENT HARRINGTON[89] at the *Relay Systems Laboratory* of the Air Force Cambridge Research Laboratory (CRL).[90] The focus of this group was the digital transmission of radar data.[91] A member of HARRINGTONS group remembers:

> "We had many troubles with radar data transmission, most of which could not be foreseen without trying the equipment out in the real world. Sending digits over telephone lines sounds easy, and it is, but sending them reliably was not. The telephone system had been elegantly designed for sending analog voice, but suffered a number of distortions and noise interferences that only digits could notice – and notice them they did. At first the telephone company was dubious about what we were doing. When the first telephone line for radar data came into the Whirlwind building to be wired into one of JACK's modems, the telephone installer insisted on wiring it into a handset. We told him we didn't want the handset, but he said it was regulations and that was that. When he left, we connected it to the modem [...]"[92]

Early studies showed that an ordinary telephone had a lower cut-off frequency of about 500 Hz and a usable bandwidth of about 1.5 kHz which could be used for long-distance transmissions.[93] This led to the development of a clever amplitude modulation scheme for data transmission. Figure 3.35 shows the block diagram of the Digital Data Transmitter (DDT) which was developed.[94]

Shown on the left are the three inputs to the transmitter: A basic timing signal with a fixed frequency of 650 Hz, a bit-serial data line, and a synchronization line. These signals are additively mixed by three resistors with a .12 : .5 : 1 ratio.[95] The output voltage of this resistor network is used to control a varistor based amplitude modulator. The timing signal is also fed to a carrier generator, which basically consists of a cathode follower generating a distorted output signal, which is fed into a high-Q filter tuned to the third harmonic of the 650 Hz timing signal, effectively yielding a 1,950 Hz car-

---

[88] Later this technology was used to connect multiple computer installations in a nationwide network.
[89] 05/04/1919–12/13/2009
[90] See [HARRINGTON 1983] for more in-depth information about the historic development of radar data transmission.
[91] Eventually, the AN/FST-2 *Coordinate Data-Transmitting Set* was developed by the Borroughs Corp. for the so-called *Cape Cod system* (see section 4). This system was, in fact, a digital computer connected to surveillance radar systems and Identification Friend or Foe (IFF) equipment (see [OGLETREE et al. 1957]). It performed the detection of potential targets, it determined range and azimuth for these targets and preprocessed these data for final transmission by telephone line.
[92] See [HARRINGTON 1983][p. 370].
[93] See [HARRINGTON et al. 1954][p. 9].
[94] See [GLOVER 1955/1] for detailed information about the DDT. It is noteworthy how much effort went into the development of marginal checking routines for all basic circuits – this becomes especially clear in [GLOVER 1955/1] and [GLOVER 1955/2].
[95] This ratio is implemented by a 820 kΩ, a 200 kΩ and a 100 kΩ resistor.

3.6 Data transmission

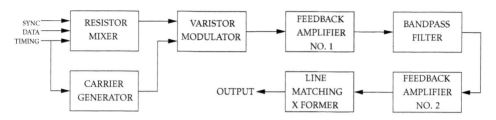

Figure 3.35: Block diagram of the DDT (see [GLOVER 1955/1][p. 1])

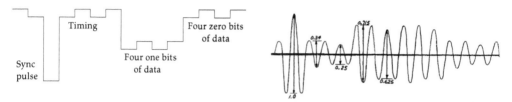

Figure 3.36: Typical output signal of the resistive mixer of the DDT (see [GLOVER 1955/1][p. 2])

Figure 3.37: Modulator output corresponding to figure 3.36 (see [GLOVER 1955/1][p. 3])

rier phase locked to the basic timing signal. Data is sent at a rate of 1,300 Hz, so that two bits are transmitted during every single oscillation of the timing signal.

The carrier and the output signal of the resistive mixer are combined in the modulator, yielding a sinusoidal output voltage which is amplified, filtered, amplified again and then fed into the telephone line by an impedance matching transformer. Figure 3.36 shows a typical output waveform of the additive resistor mixer: The start of a transmission is signaled by a sync pulse, followed by two cycles of raw timing information. Then two groups of four bits are transmitted. The corresponding output of the varistor modulator controlled by this input signal is shown in figure 3.37. This signal is fed to the telephone line after suitable amplification and filtering.

On the receiving end either a Digital Data Receiver (DDR) circuit or a Gap Filler Input Receiver (GFIR) were connected. These two circuits differed only marginally with respect to the transmission rate but were identical otherwise. Although these receivers were, as always, more complicated than the respective transmitter circuit,[96] the basic principle of operation was simple: The incoming signal was fed into an impedance matching transformer which in turn drove some amplifier stages yielding an output level of about 30 V for a synchronization pulse. The output of this so-called *signal detector* was then connected in parallel to three filter circuits: The synchronization pulse and the serial data bits were extracted by two amplitude discriminator circuits, while the timing pulses were recovered by a frequency discriminator. The three digital output signals controlled a shift register which reassembled the transmitted data words. Transmission tests performed with this equipment by transmitting 180,000 16 bit words per hour showed that the average number of errors per hour was only 6.5, of which 3.8 were readily detected by parity checks, so that only 2.7 erroneous words were left undetected.[97]

---

[96] [GLOVER 1955/2][p. 12] contains the detailed circuit schematic of a receiver.
[97] See [HARRINGTON et al. 1954][p. 18].

These modulator and demodulator devices were mainly used to transmit *Slowed Down Video (SDV)* radar data[98] to Whirlwind. The radar data was eventually stored in a 14 bit buffer drum register, corresponding to bits 2 to 15 of a Whirlwind word. The upper eight bits contained azimuth data while the lower six represented range data.[99]

3.7 Oscilloscope displays and light guns

Another novel feature of Whirlwind were the oscilloscopes attached to it. The idea of controlling the beam intensity and driving the deflection plates (in case of electrostatic deflection) or the electromagnetic deflection coils of a *Cathode Ray Tube (CRT)* was born with the invention of the WILLIAMS-KILBURN-tube[100] which required a rather intricate control system. So connecting an oscilloscope for display purposes to a digital computer like Whirlwind was a logical next step.

In 1952, Whirlwind featured two 16 inch magnetically deflected CRTs[101] of which one was directly visible while the second had a computer-controlled 35 mm Fairchild camera[102] mounted on top of its screen. The resolution in x- and y-direction was $2,048 \times 2,048$ points. The maximum rate for displaying points was 6,250 points per second. To obtain a flicker-free display, the 16 inch oscilloscopes used a long-persistence phosphor. A mere 5 μs intensification of the beam caused a visible spot on the display which persisted for about one minute.[103]

Figure 3.38 shows a typical screen shot from such an oscilloscope connected to Whirlwind. The symbols, letters and digits shown on the left are all based on 3×5 arrays of individual dots.[104] It was also possible to switch the oscilloscopes from point-mode into vector mode in which vectors, defined by start- and end-coordinates, were displayed automatically by a vector generator in about 197 μs per vector. In addition to this an interesting way to display numbers and symbols was also provided: A character generator produced a rectangular pattern of lines representing a figure 8 quite like today's seven segment displays. Under program control it was then possible to intensify selected lines from this basic figure. One such character took 557 μs to display.[105]

---

[98]The idea of SDV was to integrate the raw radar signal over several rotations of the antenna in order to remove noise etc. The resulting signal had a sufficiently narrow bandwidth to be transmitted directly, although the attainable resolution was rather low at one degree for azimuth and one mile for range data (see [HARRINGTON 1983][pp. 372 f.]).

[99]See [MUHLE 1958][p. 74 ff.]. The reduced domain for range data is in contrast to earlier experiments described in section 2.3, where range data was transmitted as seven bits.

[100]See section 3.2.1.

[101]In addition to these large screen displays, there were two slightly modified Dumont 304-H five inch oscilloscopes available (see [HEART 1952][p. 2] and [PWSR 1951][p. 6-130]). Later, when Whirlwind was used to control an experimental *Direction Center (DC)*, another set of thirty 16 inch oscilloscopes and nineteen 5 inch displays were installed in the DC (see [MUHLE 1958][p. 59]).

[102]See [HEART 1952][pp. 3. f.].

[103]See [PWSR 1951][p. 6-130].

[104]According to [MIT 1955][p. 4], approximately 320 numbers of five digits each could be displayed on a screen, taking about ix seconds to generate the display's contents.

[105]See [MIT 1955][p. 4].

3.7 Oscilloscope displays and light guns   57

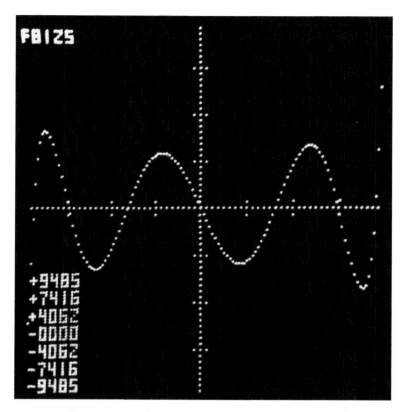

Figure 3.38: Sample oscilloscope output from Whirlwind showing a curve together with *x*-coordinates of zero crossings (see [EVERETT 1951][p. 71])

Each point, be it a single point or the start- or end-coordinate of a vector was represented by two eleven bit values. Each of these values was transmitted to a *Digital to Analog Converter (DAC)* which in turn controlled either the deflection system of the display tube directly or the vector generator. Using a so-called *scope intensification line* a particular oscilloscope could then be selected to display the point or vector by intensifying its beam.[106] The oscilloscopes could be manually connected to any of the available scope intensification lines or even a combination thereof.[107]

Another remarkable development for Whirlwind was the *light gun*, invented by ROBERT R. EVERETT,[108] which allowed a manual operator to indicate interest in a selected object or point displayed on an oscilloscope's screen.[109] At its heart the light gun contained a photomultiplier tube mounted in a gun-like enclosure which could be pointed with its photosensitive end to an arbitrary point on an oscilloscope's screen. By pushing

---

[106] See [HEART 1952][pp. 2 f.].
[107] Some programming examples can be found in [YOUNG 1954].
[108] See [TROPP et al. 1983][p. 391].
[109] The mere fact that a light *gun* instead of a light *pen* was developed, is a nice reflection of the military background of most of the research in Whirlwind.

a trigger button the computer was informed that the light gun had been pointed to a particularly interesting structure on the display. The output of the photomultiplier tube controlled a particular bit in one of Whirlwind's flip-flop registers, which was accordingly set when the selected screen area was illuminated by the electron beam of the oscilloscope. So all the software had to do was checking for such a set bit after displaying a point, character or vector.[110]

3.8 Time register

Whirlwind featured a *time register*, which allowed to count time intervals with a resolution of 250 ms. This register was actually a 15 bit counter, so that intervals up to 8,192 seconds could be measured at this resolution. A small synchronous motor connected to the mains drove an opaque disk with slots which allowed a beam of light to reach a photocell assembly every 250 ms. The amplified and pulse-shaped output of this detector was then fed to the counter circuit. Since there was no time-standard used, the time register was not suitable to derive an absolute time value. Only time intervals on which velocities of targets etc. were calculated could be determined.[111]

3.9 Reliability, power supplies and marginal checking

Reliability was one of the main objectives of Whirlwind's development. While there had been other large digital computers before, like ENIAC, their reliability was often marginal and productive work was often restricted by frequent failures, ranging from overall system failures, to subtle miscalculations which were hard to debug. There are a couple of areas which affect reliability of a machine like Whirlwind: First of all, the power supplies must deliver stable output voltages without spikes, drops, excessive noise, or hum. Another important area is that of noise in the computer's circuits. These have to exhibit a large enough signal-to-noise ratio to ensure reliable operation:

> "One important issue was our uncertainty about thermal noise. We didn't know if random spikes of thermally generated noise were big enough to trigger our robust computing circuits. We wondered whether thermal noise would intrude itself often enough to be devastating to accurate computation. To test for this, the five-digit multiplier[112] was run continuously. Every multiplication was checked against a reference number. Sure enough, it didn't compute reliably all the time. It had a great tendency to make mistakes at 3 a.m. This was traced to the janitor in the building next door, who would start the freight elevator at about that time, upsetting the power circuits enough to produce a computation error. As a result, a rotating motor generator with enough inertia to carry

---

[110]Whirlwind did not feature interrupts, so polling for events was necessary.
[111]See [HEART 1952][p. 5].
[112]This *five-digit multiplier* was a test assembly specifically built to gain experience with the design and operation of high-speed digital circuitry.

3.9 Reliability, power supplies and marginal checking

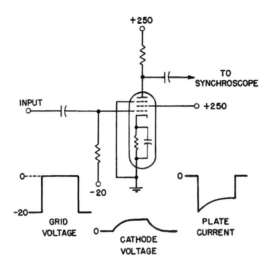

Figure 3.39: Effect of cathode poisoning (see [RICH 1950][p. 7])

through that kind of transient noise was installed on [...] Whirlwind [...] It was an expensive solution but a very effective one."[113]

The introduction of a motor-generator unit solved the main problem of supplying stable power to the computer, a technique used well into the 1980s for powering mainframes and supercomputers. The next problem to be solved, was that of tube reliability: Typical commercial tubes available in the late 1940s and early 1950s were aimed at radio and eventually television applications. The circuits employed there were of a purely analog nature, where minor degradations of the tubes can often be compensated for by cleverly designed feedback circuits. Furthermore, a typical tube was only operated for a couple of hours a day at most, while tubes for digital computers would see an active life of thousands of hours of continuous operation.

Tests performed in 1948 showed that the standard pentode 6AG7 had a life-time much shorter than that required for application in Whirlwind. Thus it was decided to switch to another model, the 7AD7 pentode. Unfortunately, this tube type also showed signs of rapid deterioration in operation, so a closer examination of the causes was necessary.[114] It turned out that evaporating silicon from the tungsten-alloy of the heater was the main culprit. Accordingly, tubes manufactured especially for Whirlwind with pure tungsten heaters showed much better behavior. This change also alleviated the problem of *cathode poisoning*[115] which is caused by running a tube in the cut-off region, i.e. with the control grid blocking the flow of electrons, for prolonged periods of time. Under these circumstances, a layer of Barium orthosilicate and other silicates forms on the cathode, effectively diminishing the amount of electrons emitted.[116] This layer

---
[113]See [FORRESTER 1988][p. 11].
[114]See [REDMOND et al. 1980][p. 92] and [REDMOND et al. 1980][p. 130].
[115]Also called *sleeping sickness*.
[116]See [RICH 1950].

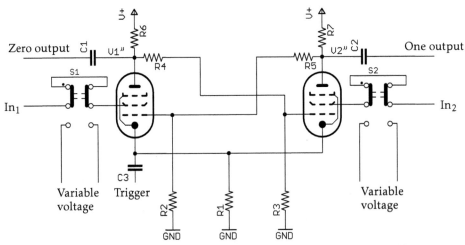

Figure 3.40: Marginal checking of a Whirlwind flip-flop (see [REDMOND et al. 1980][p. 87])

effectively forms an RC-combination in series with the cathode connection within the tube itself, as shown in figure 3.39, causing severe signal distortion.

Based on these findings, the 7AK7 pentode was developed which was used extensively in Whirlwind. In 1951 it was noted that the *"wholesale replacement of tubes within less than 20,000 hours was considered to be not only unnecessary but unwise"*.[117] In addition to all this, a considerable amount of thought was put into the development of circuits exhibiting good stability even when circuit elements changed their characteristics due to age, temperature etc.[118]

Although these measures were necessary preconditions for a stable operation of such an intricate machine as Whirlwind, the overall reliability achievable by such rather passive means would not have been enough for the proposed real-time application. Thus so-called *marginal checking* was developed. The idea was as simple as it was promising: When a circuit failed during normal operation of the computer due to aging, it would probably fail earlier when its supply and input voltages were varied to their design margins. So instead of just passively waiting for a circuit to fail, it could be pressed to fail early by applying such voltage variations during a maintenance period. Figure 3.40 shows a typical Whirlwind flip-flop with additional inputs connected to the marginal checking circuitry. In this simplified example check voltages are added only to the input signals of a flip-flop circuit while it was also common to vary plate voltages, too.

Figure 3.41 shows an early version of the *marginal checking control* panel of Whirlwind. Using the telephone dial a particular circuit to be tested under voltage excursions could be selected manually. The dial controlled the cross-bar switch shown in figure 3.42 which would select supply and signal lines for injecting marginal checking voltages. These voltages could be either varied manually or by executing a variation cycle which ran from a lower to an upper limit automatically.

---

[117] See [REDMOND et al. 1980][p. 197].
[118] See [TAYLOR 1953].

3.9 Reliability, power supplies and marginal checking

Figure 3.41: Marginal checking control panel (see [SUMNER 1950][p. 9])

Figure 3.42: Marginal checking relays (see [SUMNER 1950][p. 9])

The daily marginal checking routine was further supported by a *Consolidated Test Program*[119] which contained thirteen sub-test-programs, each testing a particular part of Whirlwind's hardware while voltages and signal levels were varied. This program not only checked the operation of the hardware but also took over control of the marginal checking subsystem by controlling the cross-bar switch and varying the voltages. All in all 399 lines could be selected for marginal checking and fast or slow cycles of voltage variation could be executed.

Table 3.1 gives a summary of part failures observed in Whirlwind during 3,054 hours of operation. Especially noteworthy is the amount of reliability achieved with the 7AK7 pentodes as compared to the 7AD7. The application of marginal checking finally made Whirlwind a truly reliable machine – far ahead of commercial machines available at that time.

Finally, it should be noted that operating a vacuum tube based computer like Whirlwind was sometimes even dangerous. Especially the high DC supply voltages for the plates posed a special risk as ORNSTEIN remembers:

> "[...] I'd encountered a fellow who had nearly been killed at Whirlwind when someone turned on the power unexpectedly while he was working with his hands in the machine. There were some high voltages in the racks and the shock he suffered had damaged him for life."[120]

---

[119] See [MORRISON 1954].
[120] See [ORNSTEIN 2002][p. 83].

Device	Type	Number in use	Failures Number	Failures Percent	Located by marginal checking Number	Located by marginal checking Percent
Tube	7AK7	1,412	18	1	2	11
Tube	7AD7	1,622	243	15	168	70
Tube	Others	1,187	92	8	20	22
	Total	4,221	353	8	190	54
Crystal	D.357	7,500	64	0.09	32	50
Crystal	D-358	3,500	278	8	197	70
Crystal	D-359	400	2	0.05	0	0
	Total	11,400	344	3	229	64
Passive	Capacitors	21,107	4			
Passive	Chokes	3,989	0			
Passive	Resistors	26,210	16			
Passive	Pulse Transformers	3,425	7			
Passive	Delay Lines	143	2			
Passive	Power Connectors	991	3			
	Total	55,865	32			

Table 3.1: Observed part failures in Whirlwind during 1950 after 3,054 hours of operation (see [REDMOND et al. 1980][pp. 138 ff.])

3.10 Programming

Programming an early digital computer like Whirlwind was different from programming today's machines in several respects. Some of the most obvious differences are the tiny amount of main memory, the low speed of execution etc. But more important from a programmer's perspective is the different programming model itself, where subroutine handling is done by modifying instructions in the program itself, rather than relying on a stack data structure held in memory. In addition to this, most of the early programming was done in pure octal since no support programs like assemblers were available in the beginning. When assemblers, which not only allowed the use of so-called *mnemonics* instead of hard-coded operation codes but also alleviated the programmer from the tedious and error-prone task of manual address calculation, finally emerged, there was considerable opposition among some of the programmers:

> "A study of methods of programming with the present and proposed conversion programs shows that most programs, with present methods, can be written in such a fashion that they can be assembled automatically, with no calculations of storage positions on the part of the programmer. [...] The philosophy of this memorandum is diametrically opposed to that of the programmer who uses octal notation, likes it, and is convinced that this is the most satisfactory way to write a program. Nevertheless, under certain conditions, the method proposed here may prove to be a much more speedy, more easily corrected way to give complete machine instructions."[121]

---

[121] See [CARR 1952][p. 1].

3.10 Programming

Instruction	Effect
1. ca 201	
2. ts 203	
3. ca 200	a
4. dv 203	a/x_n in BR
5. sl 15	a/x_n in AC
6. su 203	$a/x_n - x_n$
7. sr 1	$(a/x_n - x_n)/2 = x_{n+1} - x_n$
8. ts 204	
9. ad 203	x_{n+1}
10. rs 203	
11. cm 204	$\|x_n - x_{n+1}\|$
12. su 202	
13. cp next job	
14. sp 3	Return for calculation of next x_n

Data

200. a	
201. $x_1 = 1 - 2^{-15}$	
202. 2^{-14}	
203. _ _ _	Used for x_n, finally $\sqrt{a}$
204. _ _ _	Used for $\left(\frac{a}{x_n} - x_n\right)/2$

Figure 3.43: Computing $\sqrt{x}$ by applying NEWTON's iterative method (see [SAXENIAN 1951][p. 47])

Figure 3.43 shows a simple example program for Whirlwind.[122] It computes an approximation to $\sqrt{a}$ by NEWTON's[123] iterative method and is based on the iteration

$$x_{n+1} = x_n + \frac{\frac{a}{x_n} - x_n}{2}.$$

The value a, the initial iteration value, $x_1 = 1 - 2^{-15}$, and the desired accuracy $\varepsilon = 2^{-14}$ of the solution are initially stored in memory locations 200, 201, and 202. The first instruction, ca 201, short for clear and add, clears the AC and adds the content of memory location 201, effectively loading the AC with x_1. The next instruction prepares the first iteration loop by executing a transfer to storage instruction ts 203 which deposits the contents of the accumulator in the memory location 203.

The main loop consists of the instructions starting at address 3 and ending at address 14. ca 200 loads a into the accumulator, and dv 203 divides it by x_n which has been stored previously into location 203. The result of this division, which is still in BR,

---
[122]See [SAXENIAN 1951] which contains a wealth of practical examples of Whirlwind programming. Details on programming Whirlwind can be found in [SAXENIAN 1951], [MUHLE 1958], [HEART 1952], [BAGLEY et al. 1954], [MIT 1951], [LONE 1952], and [Whirlwind Programming Notes].
[123]12/25/1642–03/20/1726

is transferred to the AC by a 15 bit left-shift, caused by sl 15. su 203 now subtracts x_n from this value, leaving $a/x_n - x_n$ in the accumulator. Division by two is then accomplished by shifting the accumulator's contents one bit to the right with sr 1. The result, $(a/x_n - x_n)/2$, is then transferred to storage address 204 by ts 204.

Adding the contents of memory location 203 by ad 203 yields x_{n+1} which is stored back with ts 203. The clear and add magnitude instruction cm 204 loads the absolute value of memory location 204 into the accumulator. Subtracting ε prepares the following conditional program instruction. If the contents of the AC are negative, cp next job will store the contents of the program counter plus one in the AR and load the Program Counter (PC) with the destination address. If the contents of the AC are non-negative, the cp instruction is skipped and sp 3, short for sub program, is executed, which loads the program counter with the absolute address 3.[124]

Programming Whirlwind took place in a rather shirt-sleeved atmosphere which had next to nothing in common with the commercial closed-shop systems where programmers delivered anonymous batches of punch cards to a computer installation where they were processed with results printed out on line-printers, being delivered hours or even days later to the programmer:

> "When the computer was first made available to the applications group, each programmer operated his own program. When alarms occurred it was common to see the computer sit idle for large periods of time while the programmer desperately tried to guess what the trouble was. Usually he would randomly examine the contents of some storage registers or operate the computer in an 'order-by-order' fashion to trace the path of control. Not infrequently a wrong button pressed or a switch forgotten, under the pressure of the moment, resulted in destruction of the symptoms he was trying to diagnose. The majority of these situations ended with the next programmer, his patience at an end, demanding his turn and the woeful programmer, still puzzled, returning to his office."[125]

Whirlwind was not just programmed in machine language or assembler – a programming system called the *Comprehensive System* was developed in the early 1950s. This allowed the input of programs in alphanumeric form using mnemonics like those shown in figure 3.43, and the automatic calculation of addresses. In addition to this, it implemented an early form of a virtual machine. This virtual machine interpreted instructions for a hypothetical machine on Whirlwind and supported floating-point arithmetic among other features:

> "[It p]ermits the intermingling of machine code with interpreted code, yielding either 35,000 4.5-decimal-digital fixed-point operations per second or 800 7-decimal-digital floating-point operations per second. All Whirlwind input, output and secondary storage units are available through automatically-selected subroutines or direct machine code as desired. Changing from interpreted to machine code and vice versa is accomplished by jump instructions with special

---

[124] A list of all Whirlwind instructions can be found in appendix A.
[125] See [GILMORE 1951][p. 2].

3.10 Programming

addresses, said instructions being designated by the words IN and OUT, for into and out of interpreted code."[126]

Another system developed for Whirlwind was called the *Summer Session Computer* and was designed with the focus on its application in classrooms for teaching computer programming. It was much simpler to use than the Comprehensive System but this came at a price: Programs written for the Summer Session Computer ran considerably slower since it implemented an interpretive system only without any provisions for switching forth and back between interpreted and native machine instructions.[127]

As impressive as these early virtual machines are, there was an even more impressive development – the *Algebraic System* developed for Whirlwind by J. HALCOMBE LANING[128] and NEAL ZIERLER.[129] This system, which became informally known as *George* after the radio drama series "Let George Do It" which aired between 1946 and 1954, was completed in 1954.[130] Figure 3.44 shows a simple George example program. It evaluates a function $z(x, y)$ for $x = 173.972$ with y being an integer value running from 15 to 30.[131]

```
x=173.972,
y=15,
1   z=x+y²/(7x-y(x³+7+3x⁻³)¹⁵),
    PRINT y,z,
    y=y+1,
    c=y-30.5,
    CP1,
    STOP,
```

Figure 3.44: Algebraic example program (see [ADAMS 1954][p. 21])

Apart from these early and successful forays into the development of programming languages, other seminal ideas also emerged from the Whirlwind project. Among these is the idea of shareable libraries. Another early abstract idea of connecting two computers to form a machine which could run a *multi-sequence program*[132] antedates later parallel processors and programming models for these and led to the concept of multi-threading:

---

[126] See [ADAMS 1954][p. 7].
[127] For more information on the Summer Session Computer see [ADAMS et al. 1954].
[128] 02/14/1920–05/29/2012
[129] 1926–
[130] See [LANING et al. 1954] for more information.
[131] Although George may not have influenced the development of FORTRAN which took place in parallel at IBM, the Algebraic System is one of the earliest programming languages featuring a "natural" input format, hiding the machine details from the programmer.
[132] See [CLARK 1954].

> "A multi-sequence program can also be constructed for a single computer. The general requirement is that the operation of one sequence must not interfere with the operation of any other. In general, this means that the operating registers of the computer must be time-shared by the sequences. A program counter must be provided for each sequence and the computing system must include an element for deciding which sequence is to be advanced during any given control cycle."[133]

All in all, Whirlwind not only demonstrated that a large scale computer was feasible with the technologies of the 1950s, successfully showed the application of a digital computer for real-time control applications, paved the way for complex input/output equipment and graphical user interfaces, but also pioneered many of today's software technologies.

3.11 The end of Whirlwind

After its development and its successful application to solve a wide variety of problems, responsibility for Whirlwind was transferred to Lincoln Laboratory in 1957, where the machine was used for further work in the air defense program. The machine was finally shut down on May 17th, 1959 after *"62,000 filament hours"*[134] but this turned out to be not the end yet.

Whirlwind was leased by an alumnus of the project, WILLIAM WOLF,[135] then president of the "Wolf Research and Development Corporation" in 1959 which was confirmed by the Office of Naval Research.[136] The system was finally put back into operation in 1963. The process of dismantling, moving and reassembling Whirlwind took nearly a year and was planned using a *PERT*[137] network, then still a novelty. Just the task of reconnecting all cables and wires of the machine ran from April 30th, 1962 to October 29th, 1962 showing the immense complexity of this undertaking.[138]

From today's perspective, the operation of a large vacuum tube based computer seems adventurous and hazardous as the following quotation shows:

> "[...] it was noticed that power was being applied to a relay associated with the magnetic tape racks. When the filament switch for the tape racks was switched to what was thought to be the off position to turn off this relay, two No. 10 filament wires suddenly burned up. When this switch was thrown, it effectively placed a short across the output of the 400 amp standby alternator, due to some temporary jumpers which had been installed last summer and which

---

[133] See [CLARK 1954][p. 2].
[134] See [REDMOND et al. 1980][p. 224].
[135] *1928
[136] See [WILDES et al. 1986][p. 338] and [The Tech][p. 1].
[137] Short for *Program Evaluation and Review Technique*.
[138] See [SHORTELL 1963] and [SHORTELL 1964] for a highly interesting and thorough description of the checkout procedures and progress of Whirlwind at its new location.

3.11 The end of Whirlwind

had not been removed. This sudden overload on the alternator burned up these wires, rather than blowing the 600 amp fuse in the output of the alternator."[139]

Even supporting equipment like air conditioning turned out to be neither easy to operate nor to maintain:

"On Monday, June 17, exactly one week after instituting twentyfour hour operation [of the air conditioning equipment], we arrived to find the system shut down and the floor covered with oil. Sometime during the week-end the number 4 compressor had blown a flare fitting in an oil line which resulted in complete loss of oil in the entire system and very possibly simultaneous loss of a few cylinders of Freon."[140]

Despite these difficulties, Whirlwind ran successfully until 1973 when it was finally shut down for good. In January 1976 some of its central parts were donated to the Smithsonian Institution; other parts were on display at the Digital Equipment Computer Museum.[141] The latter parts, including the two first magnetic core memory subsystems are now part of the collection of the Computer History Museum. These parts and many fond memories are all what exists of Whirlwind today.

---

[139] See [SHORTELL 1963][p. 36].
[140] See [SHORTELL 1963][p. 37].
[141] See [WILDES et al. 1986][p. 338] and [REDMOND et al. 1980][p. 224].

4 SAGE

After Whirlwind had successfully demonstrated the feasibility of a high-speed digital computer for real-time control in general and its successful application to tracking air targets, it became clear that only a machine like this could be the key to an effective solution of the air defense problem which had been described by EVERETT as follows:[1]

> "The air-defense data-processing problem is one of nationwide data-handling capability: facilities for communication, filtering, storage, control, and display. A system is required that can maintain a complete, up-to-date picture of the air and ground situation over wide areas of the country; that can control modern weapons rapidly and accurately; and that can present filtered pictures of the air and weapons situations to the air force personnel who conduct the air battle."[2]

These requirements were further aggravated by the peculiarities of a possible attack with nuclear weapons carried by long-range bombers. Such configurations were no longer one-of-a-kind weapons but actual weapon systems which had been deployed in large numbers in the United States of America as well as in the Soviet Union, resulting in an increasing threat to the world's overall safety:

> "Nuclear war presented Air Force officers with stark options that had to be executed with little or no deliberation. Operational commanders had to determine in a matter of minutes whether the appearance of unknown aircraft or missiles on a warning system device constituted a nuclear strike or rather were phenomena related to atmospheric conditions or technical problems in the warning system itself. Based on that assessment, commanders then had a minute or two in which to order a nuclear counterstrike with bombers and later with missiles. A mistake either way – launching a strike erroneously or failing to launch the counterstrike quickly in case of an actual attack – would lead to the destruction of the United States. In such circumstances, Air Force leaders wanted every extra second to make the proper decisions. They also demanded that their information be reliable and then that their commands be executed rapidly."[3]

Since a single computer like Whirlwind would have been neither powerful nor robust enough from a military point of view, a network of computers had to be built, something neither the MIT, nor the Lincoln laboratory ever had done or could be expected

---

[1] A thorough description about the historical background of the air defense problem, which has been characterized quite aptly as *"Too much data and not enough information."* (see [BOSLAUGH 2003][p. 62]), may be found in [ASTRAHAN et al. 1983][pp. 341 ff.].
[2] See [EVERETT et al. 1983][p. 331].
[3] See [JOHNSON 2002][pp. 117 f.].

to do.[4] The computers this network would be based on also had to be designed and built and would be based architecturally on Whirlwind. Thus the prototype design was initially known as *Whirlwind II*.[5] Three of the largest and best-known computer manufacturers of the 1950s were taken into consideration for this vast development program, namely IBM, Remington Rand and Raytheon.[6] Eventually, IBM was chosen as the main contractor for building the computer systems which would be at the heart of SAGE and would finally become known as AN/FSQ-7. The cooperation between the MIT based group and IBM was not easy at the very beginning:[7]

> "With the considerable experience they had gained in designing Whirlwind and its application to air defense, FORRESTER's Division 6 engineers had no intention of letting IBM dominate the design of the AN/FSQ-7. In fact, they planned to design the computer and then hand over the design to IBM engineers, who might make minor modifications for production purposes. By contrast, IBM's engineers were fresh from the design of the highly successful 701 computer and considered themselves the preeminent experts in computer design. They expected to design and deliver the AN/FSQ-7 as they had always done in the past. In practice, two headstrong and highly competent organizations would butt heads and have to discover how to work together and learn from each other. Their first meetings were 'loud and rancorous' but eventually the teams came to respect each other's abilities."[8]

Initially, two prototype systems, called *XD-1* and *XD-2*, short for *experimental development*, were built. XD-1 was installed at Lexington in 1955 where it became the heart of the *experimental SAGE subsector* which will be described in the following, while XD-2 found its home at IBM's Poughkeepsie plant where it was primarily used as a testbed for ongoing hardware developments and for training purposes.

---

[4] "Because the [Lincoln] Laboratory's role as an MIT research and development organization did not extend to system implement, in 1958 some personnel from Lincoln Laboratory left to form the MITRE [see [MURPHY 1972] and [MITRE 2008]] Corporation to complete the engineering for SAGE deployment." (see [GROMETSTEIN 2011][p. xiii]).

[5] "Whirlwind I was more of a breadboard than a prototype of a computer that could be used in the air defense system. The Whirlwind II group dealt with a wide range of design questions, including whether transistors were ready for large scale employment (they were not) and whether the magnetic-core memory was ready for exploitation as a system component (it was). The most important goal for Whirlwind II was that there should be no more than a few hours of down time per year." See [GROMETSTEIN 2011][p. 22].

[6] See [GROMETSTEIN 2011][pp. 22 ff.] and [ASTRAHAN et al. 1983][pp. 343 ff.] for more details on vendor selection. Incidentally, although it had no influence on the final decision, IBM already had some connections with Whirlwind since NATHANIEL ROCHESTER (01/14/1919–06/08/2001), the chief architect of the IBM 701 computer, the so-called *defense calculator*, had built the arithmetic unit of Whirlwind back in 1947 (see [ROCHESTER 1983][p. 115]).

[7] One of the reasons why this cooperation worked rather well in the end may have been due to the fact that IBM's management was well aware of the utmost importance of this particular project to the company's future: "IBM will be recognized as the undisputed leader in the large scale, high-speed, general purpose digital computer field. If a competitor were performing on this contract, that competitor might gain enough advantage to force IBM into a relatively secondary position." (see [GREEN 2010][pp. 192 f.]).

[8] See [JOHNSON 2002][p. 136].

4.1 The Cape Cod system

In 1951 it was decided to setup a functional prototype of the later air defense system for the evaluation of various new technologies which would be necessary to accomplish the overall task in large scale. This system, called the *Cape Cod system* due to its location, was developed and set up by the Lincoln Laboratory and became operational in September 1953.[9] At its heart was still Whirlwind which had been extended with regard to drum storage and input/output equipment and had been coupled with a long-range radar system and several Gap Filler (GF) radar systems by means of digital data transmission over telephone lines.[10]

One of the developments resulting from this experimental system was that of the so-called *direction center*, DC for short. From this location the military personnel from the air force performed the tests, monitored the air space of the sector under surveillance and controlled the overall system. The first incarnation of this Cape Cod system, operational from 1953 on, was mainly used to gather preliminary data which would be used for the design and planning of the envisioned large scale SAGE air defense system:

> "Emphasis was directed toward singling out obvious problem areas and attempting to correct whatever difficulties were encountered, rather than toward gathering complete statistical data on system operation. Consequently there was very little modification of equipment."[11]

Based on this early system, the *1954 Cape Cod system* was set up. It used a significantly expanded network of radar stations generating input data and was finally superseded by the so-called *experimental SAGE subsector*, located in Lexington, Massachusetts. While the two predecessor Cape Cod systems relied on Whirlwind as their central computer system,[12] this subsector was already based on the XD-1, which had been built at IBM's Poughkeepsie plant, where it was dismantled and transported to its final destination at Lexington, where it was installed in January 1955.[13] In addition to the previous test-systems, this experimental subsector was equipped with a ground-to-air data link. The purpose of this radio link was to perform experiments with aircraft equipped with suitable receiving equipment. These experiments turned out to be highly successful as WIESER remembers:

---

[9][ASTRAHAN et al. 1983][pp. 342 f.] and [WIESER 1983] give a thorough account of the Cape Cod system. BENINGTON compared the Cape Cod Systems and SAGE as being *"model shop"* and *"production"* respectively (see [TROPP et al. 1983][p. 387]).

[10]It should be noted that Whirlwind was not the only digital computer under development for military purposes at that time. Other institutions also developed stored program digital computers for military purposes like the Canadian Navy's *Digital Automated Tracking and Resolving System*, or some early digital experiments at the Navy Electronics Laboratory which yielded the *Semi-Automatic Digital Analyzer and Computer*, SADZAC for short, and the *Semi-Automatic Air Intercept Control System*, SAAICS for short (see [BOSLAUGH 2003][pp. 62 ff.]).

[11]See [IBM INTRO][p. 2].

[12]Even the MTC was used for data processing.

[13]See [LUNDBERG 1955][p. 2]. Rather remarkably, XD-1 was up and running again with *"the same reliability and margins observed prior to shipment"* from IBM's Poughkeepsie plant to Lexington within three weeks after its disassembly (see [WILDES et al. 1986][p. 299]).

> "The first ground-air data link experiments were interesting. Doc DRAPER[14] of the Instrumentation Lab had a light test facility out at one end of Hanscom Field. CHIP[15] COLLINS, his chief pilot, discovered that one of the aircraft, a World War II B-26, Martin Marauder, had an autopilot that could take digital input. The radio frequencies were set up to send vectoring instructions directly to the autopilot. On the test we heard CHIP COLLINS say, 'Let George do it', which meant switch to autopilot. A little while later, when we traced it on the scopes, he said, 'Tallyho', as he sighted the target. Someone dubbed that 'The Immaculate Interception'."[16]

Other, more mundane tests were equally promising – a *System Operation Test (SOT)* performed between November 1955 and January 1956, consisting of eight simulated attacks flown by B-29 bombers headed for Boston, showed that of 50 attacking airplanes, 24 were successfully intercepted by the system. The next SOT was even

> "more dramatic and realistic. As many as 32 high-speed B-47 aircraft flown by the Strategic Air Command attacked targets in Boston as well as Martha's Vineyard, Massachusetts, and Portsmouth, New Hampshire. Raids included multiple aircraft performing complex maneuvers: flying less than 1,000 ft apart, making turns, and crossing, splitting, or joining tracks."[17]

Eventually, more than 5,000 sorties were flown against the Cape Cod system and it finally consisted of 30 *operational stations*, each based on a large-screen *Situation Display (SD)* which displayed targets and tracks. Accompanying information was displayed by additional auxiliary displays showing numeric and text data regarding selected targets and tracks.[18] Figure 4.1 gives an impression of the size and complexity of this system. On the left is the control console of the XD-1 computer and a situation display can be seen on the right. Compared with what was to come with the final AN/FSQ-7 installations for SAGE, this was nevertheless just a test-bed system.

4.2 SAGE and AN/FSQ-7

An IBM publication nicely summarizes the role of the central AN/FSQ-7 computer in SAGE which is also depicted in figure 4.2:[19]

> "Heart of the SAGE system, is an IBM-built computer called the AN/FSQ-7. This computer digests radar returns from all sources, plus ground observer reports, flight plans and weather information.[20] It translates this information into an over-all picture of the air situation.

---

[14]CHARLES STARK DRAPER, 10/02/1901–07/25/1987.
[15]Charles
[16]See [WIESER 1988][p. 16].
[17]See http://www.ll.mit.edu/about/History/capecodprototype-2.html, retrieved 12/23/2013.
[18]See [ASTRAHAN et al. 1983][p. 343].
[19][WAINSTEIN et al. 1975][pp. 209 ff.] contains a thorough description of the inception of SAGE. See also [JOHNSON 2002][pp. 117 ff.] for a thorough description of the development of SAGE.
[20]This information had to be entered manually.

4.2 SAGE and AN/FSQ-7

Figure 4.1: Console of the XD-1 computer (see [JOHNSON 2002][p. 137] and [IBM CCS XD][p. 363])

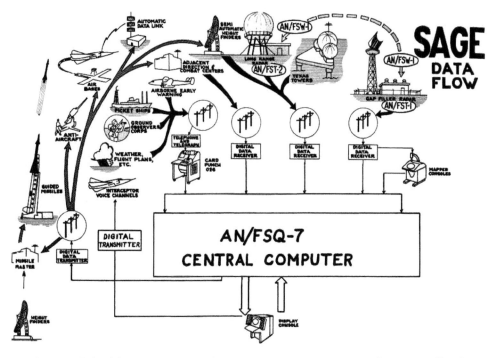

Figure 4.2: Role of the AN/FSQ-7 computer system in the SAGE system (see [IBM INTRO][p. 4])

> *The computer automatically calculates for the air commander the most effective employment of such defensive weapons as guided missiles, anti-aircraft batteries and jet interceptors. In the case of intercepting jets, the aircraft are controlled by directions fed by radio directly from the computer to the automatic pilots in the planes."*[21]

A typical sequence of events starting with the acquisition of a new target and ending at its interception looked like this:[22] A potential target would be detected by an AN/FPS-20 radar system connected to an AN/FST-2 coordinate transmitting set which was in effect a digital computer, too.[23] The AN/FST-2 computer would then combine radar data and data obtained from IFF systems if present, determine range and azimuth of the potential target and transmit these data to the AN/FSQ-7 system it was connected to. These operations were known as the *fine-grain data* function, *FGD* for short. In addition to this, the AN/FST-2 was responsible for performing a second function, the so-called *semi-automatic height finder* function (*SAH/F*). This operation was semi-automatic in the sense that a command had to be sent from the AN/FSQ-7 to the AN/FST-2 which in turn steered a height-finder radar like the AN/FPS-6 to obtain the height information of the target selected.

Based on the coordinate data received from the AN/FST-2 system, the first task performed by the AN/FSQ-7's was to decide whether these coordinates belonged to a track already known or if they represented a newly acquired target. Data belonging to an already known track was just used to update this track's information. Otherwise the coordinate set was called an *initial pickup* and a correlation process was commenced during which incoming radar data were checked against this initial coordinate set in five second intervals to determine a correlation for the initial pickup. If a correlation could be successfully established, the newly established track was marked as *tentative* and shown with additional information like speed and heading on a display console.

In parallel, the AN/FST-2 was instructed to determine a height reading on this target which was then displayed, too. During all these operations, a continuous smoothing of all current track data was performed during which incoming raw radar data was combined with predicted track data. In addition to this, the newly acquired track was then marked either *friendly* or *hostile* based on information about the current air situation stored in the memory system of AN/FSQ-7 and data returned from IFF equipment if applicable. This information was then shown on a display console together with a figure of *merit* denoting the data quality. This figure of merit was displayed as either G, F, or P denoting *good, fair,* or *poor* data quality.

In parallel to all these tasks which had to be performed for each individual target in real-time, the AN/FSQ-7 system requested status information of connected missile and interceptor sites in the background and computed preliminary estimations of intercept coordinates for targets. Based on these computations, an intercept operation

---

[21] See [IBM BOMARC][p. 11].

[22] See [IBM BOMARC][pp. 39 ff.].

[23] See [OGLETREE et al. 1957] for a more detailed description of the AN/FST-2 system. A typical duplex AN/FST-2 installation, consisting of two computers, normally one active and one in standby mode, occupied 21 cabinets, containing about 6,900 vacuum tubes and 24,000 diodes. The AN/FST-2 computers were designed and built by the Burroughs Corporation (see [The Command Post 1958][p. 3]).

could be started by pressing a *fire button* on an operator console. There was no way of starting an intercept operation without human intervention in SAGE – as General LeMay[24] put it:

> "SAGE does not think. It gathers and stores information on which man can act. SAGE does not nullify the requirement for well-trained and proficient personnel. It enables such personnel to do a better job."[25]

If an intercept operation was started, guidance information was transmitted automatically to the selected missile or interceptor. The predicted flight path of the missile or interceptor was then displayed and updated in real-time on the operator's console. The first overall system test took place on August 7, 1958, when

> "the IBM/SAGE computer at Kingston undertook the first remote-controlled intercept of a drone target by a BOMARC missile.[26] The BOMARC was fired from Cape Canaveral and the intercept was made at sea. This was the first 'live' compatibility test of the SAGE/BOMARC system."[27]

Piloted aircraft like the Convair F-106 Delta Dart interceptor could also be controlled by the AN/FSQ-7 as had been demonstrated with the XD-1 based system earlier. Patrick McGee, SMSgt, USAF (ret.) remembers:

> "The F-106A operated in conjunction with the SAGE [...] network linked via the Hughes MA-1 fire-control system to the F-106. It operated by plotting the course needed to intercept an enemy aircraft, automatically sighted the target, fired the air-to-air missiles, and then automatically placed the F-106 on the correct course to disengage. The F-106 could actually be fully computer-flown during most of its mission, the pilot being needed only for takeoff, landing, or in case something went wrong with the automation."[28]

4.3 Control and direction centers

These capabilities alone would have been remarkable enough but the overall SAGE system was much more complicated since it consisted of 23 so-called *direction centers* (DC), each based on a duplex AN/FSQ-7 installation where during normal operations one computer was active while the second one was in standby, undergoing maintenance, etc. These DCs, operating on a sector level, were connected with each other

---

[24] Curtis Emerson LeMay, 11/15/1906–01/01/1990
[25] See [NYADS 1960].
[26] BOMARC, short for *Boeing and Michigan Aeronautic Research Center*, was a supersonic – about Mach 2.7 – ground-to-air missile developed in the 1950s. It was capable of carrying either a traditional or a nuclear warhead which was triggered by a proximity fuse. The first BOMARC installation became operational in 1959, see [Lombardi 2007] or [Flight 1957].
[27] See [IBM BOMARC][p. 15].
[28] See http://www.f-106deltadart.com/history.htm#MA-1_System, retrieved 12/14/2013.

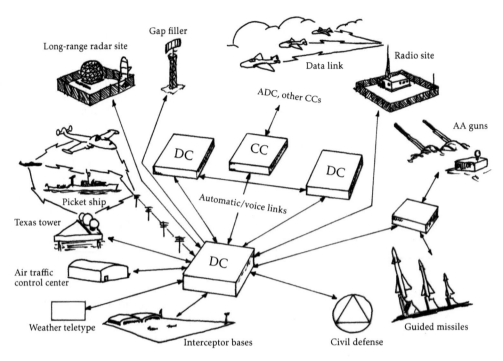

Figure 4.3: Relationship of direction and control centers (see [IBM INTRO][p. 4])

by means of digital data transmission over leased telephone lines. In addition to this, several *Control Center (CC)* installations, operating on a division level, were also part of this nationwide network. Figure 4.3 shows the relationship of DCs and CCs in the SAGE network. The role of a direction center can be described as follows:

> *"The direction center communicates with over 100 adjacent installations. Air-surveillance data are received from several types of radars. Long-range search and gap-filler radars located throughout the sector provide multiple coverage of the air volume within the sector; picket ships, airborne early warning (AEW), and Texas towers extend this coverage well beyond the coastline; height finders supply altitude data. [...] Other inputs to the direction center include missile, weapons, and airbase status; weather data; and flight plans of expected friendly air activity."*[29]

Initially, 32 such direction centers were planned, but only 23 were actually built and installed due to budget cuts.[30] For some time three CCs were also part of the network, each running a duplex AN/FSQ-8[31] installation in contrast to the AN/FSQ-7 systems

---

[29] See [EVERETT et al. 1983][p. 333].

[30] 22 DCs were located in the United States while one was installed in Canada.

[31] The AN/FSQ-8 was similar to the AN/FSQ-7 but lacked most of the former's sophisticated input/output capabilities and featured a smaller main memory system. The remaining text will focus on the larger AN/FSQ-7 on which the AN/FSQ-8 was based.

4.3 Control and direction centers

of the DCs. These AN/FSQ-8 systems performed mainly supervisory functions and had no direct radar inputs etc. Table 4.1 lists all DC and CC sites built.[32]

All in all 56 AN/FSQ-7 and AN/FSQ-8 systems were built and deployed, including XD-1 and XD-2. Each of these systems was capable of performing about 75,000 instructions per second and cost about $ 42,000,000.[33] An AN/FSQ-8 had two core memory systems of 4 k words each while an AN/FSQ-7 had one 4 k and one 64 k core stack at a word length of 32 bits. A typical installation required about 1 megawatt[34] of electrical power to run the central computer with its about 55,000 vacuum tubes, 175,000 diodes and eventually even 13,000 transistors (all of which were spread over more than 7,000 pluggable units), weighing in excess of 275 tons, and occupying about 21,000 square feet of floor space.[35]

Typically, an AN/FSQ-7 was housed in a special *blockhouse* built from reinforced concrete[36] four stories high due to the immense number of display consoles and other equipment unique to a DC, while an AN/FSQ-8 installation usually fit into a three story blockhouse.[37] A typical four-story blockhouse was 74 feet in height, containing 90,000 square feet of floor space while the smaller, three story variant used for an AN/FSQ-8-installation, was 49 feet high and contained 67,500 square feet of floorspace.[38] Figure 4.4 shows the blockhouse at McChord AFB during construction while figure 4.5 shows an aerial picture of the completed DC blockhouse at Topsham Air Force station taken around 1958. The dominating blockhouse structure can be seen on the far left bottom: The multi-story building contained the duplexed AN/FSQ-7, its associated input/output equipment, literally hundreds of display consoles[39] and the necessary air conditioning equipment,[40] telephone switching systems, etc. The flat *power house* building, 28 feet in height with 22,000 square feet of floor space, attached to the right of the blockhouse, contained five huge diesel generators delivering 650 kW of electrical energy each, powering the DC. These generators consumed about 1,800,000 gallons of diesel oil per year.[41] Powering a DC or CC from the utility power

---

[32] Compilation of this table was difficult due to a lot of contradictory data which can be found. Accordingly, the dates should not be taken for granted but as a best-guess. Some of the installations were especially remarkable: A fourth CC was planned to be installed at the Minot site and the blockhouse for the AN/FSQ-8 was built, but the computer was never installed. The installation at Sioux City is especially noteworthy as it was initially based on an AN/FSQ-8 which was subsequently upgraded to an AN/FSQ-7 by adding all of the necessary input and output equipment. The installation at Luke Air Force Base (AFB) eventually became the programming center for all other SAGE sites.
[33] See [SHULMAN 1959].
[34] [The Command Post 1958][p. 3] states that the computer's power consumption was 850 kW.
[35] See [BELL 1983][p. 13].
[36] See [The New York Times 2001].
[37] This holds true for the early layout of the blockhouses. Later AN/FSQ-7 installations (Adair AFS, K. I. Sawyer AFB, Stead AFB, Norton AFB, Beale AFB, Minot AFB, Malmstrom AFB, Luke AFB, and Sioux City AFS) were based on a a three-story blockhouse with different layout covering a larger surface area. The Canadian installation in North Bay was not based on a blockhouse architecture at all since this was the only underground SAGE installation (see [Edmonton Journal 1960]).
[38] See [Chicago Air Defense Sector][p. 2].
[39] One such display console cost $ 25,000, see [Chicago Air Defense Sector][p. 1].
[40] A typical installation required an 800 ton air conditioner (see [The Command Post 1958][p. 3]).
[41] See [Chicago Air Defense Sector][p. 2].

Sector	DC	CC	Location	State	Activation	Deactivation
New York	01		McGuire AFB	NJ	07/01/1958	09/30/1968
Boston	02		Stewart AFB	NY	01/08/1957	12/31/1969
Syracuse	03		Hancock Field	NY	01/01/1959	09/23/1983
Washington	04	01	Fort Lee Air Force Station (AFS)	VA	01/08/1957	03/01/1983
Bangor	05		Topsham AFS	ME	03/01/1959	09/30/1969
Detroit	06		Fort Custer	MI	09/01/1959	09/30/1969
Chicago	07	02	Truax Field	WI	10/01/1959	12/31/1967
Kansas City	08		Richards-Gebaur AFB	MO	1961	12/31/1969
Montgomery	09		Gunter AFB	AL	07/01/1958	12/31/1969
Duluth	10		Duluth International Airport (IAP)	MN	11/15/1959	04/01/1966
Grand Forks	11		Grand Forks AFB	ND	11/15/1959	12/01/1963
Seattle	12	03	McChord AFB	WA	07/01/1958	08/04/1983
Portland	13		Adair AFS	OR	09/01/1958	09/30/1969
Sault Sainte Marie	14		K. I. Sawyer AFB	MI	06/15/1960	12/15/1963
Spokane	15		Larson AFB	WA	09/08/1958	09/01/1963
Reno	16		Stead AFB	NV	02/15/1959	04/01/1966
Los Angeles	17		Norton AFB	CA	02/01/1959	06/25/1966
San Francisco	18		Beale AFB	CA	02/15/1959	08/01/1963
Minot	19		Minot AFB	ND	1958	08/15/1963
Great Falls	20		Malmstrom AFB	MT	02/15/1960	03/01/1983
Phoenix	21		Luke AFB	AZ	06/15/1959	12/09/1983
Sioux City	22		Sioux City AFS	IA	12/01/1961	04/01/1966
Goose (Canada)	31		Canadian Forces Base (CFB) North Bay	ON	1963	07/01/1983

Table 4.1: SAGE DC, CC and SCC installations

4.3 Control and direction centers   79

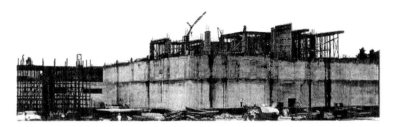

Figure 4.4: Blockhouse at McChord AFB during construction (see [Davison 1977][p. 10])

Figure 4.5: DC building with power house (lower left) at Topsham Air Force Station, Maine, ca. 1958 (courtesy Air Force Historical Research Agency)

grid alone was considered being too error-prone, thus these installations relied heavily on their local diesel engines and generators.

Figure 4.6 shows the internal structure of a typical DC blockhouse: The ground floor was occupied with air conditioning systems[42] and telephone equipment with its associated power supplies. The second floor housed the huge duplex AN/FSQ-7 installation, while the third floor contained offices and the subsector command post featuring a projecting unit which allowed display of the current air situation on a very large screen. The fourth floor finally housed a plethora of more than one hundred sophis-

---

[42]*'The air temperature entering the computer is approximately 60 degrees [Fahrenheit]. The exhausted air is seventy seven degrees. The AN/FSQ-7 computer system has four independent air conditioning systems. [...] The computer is expected to be capable of continuing operation until the temperature reached 100 to 120 degrees.'*, see [CEM 55-19][Atch. 11].

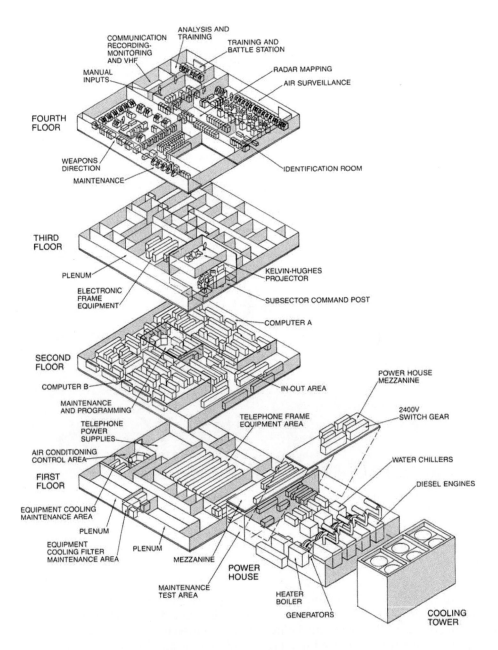

Figure 4.6: Floor plan of a SAGE direction center

Task	Maximum	Minimum
Scope operators	165	100
Officers in command	35	30
Technicians	25	10
Building maintenance	15	7
Diesel mechanics	15	3
Telecommunications maintenance	20	10
Security	30	25
Food preparation	30	20
Miscellaneous	15	4
Total for 3 shifts	990	627

Table 4.2: Staffing of a typical SAGE site

ticated display consoles and other input/output equipment. These were arranged in several groups – each assigned to a different task: The *air surveillance* consoles were used to monitor the air defense situation of the sector controlled by this particular DC. The *radar mapping* room housed one console for each gap filler radar connected to the installation – covering areas which were not of interest since they generated ground clutter or the like, could be covered on the screens with a grease pencil so that only *blips* of interest showed through and could be picked up by photomultiplier tubes mounted on top of the display consoles. Manual input devices like Teletypes as well as card punch and reading equipment were grouped in the *manual input* room, while the *weapons direction* room contained the consoles manned by the *senior weapons director* who assigned tracks to be intercepted to *weapons direction teams* who in turn controlled the intercept operations.[43] Additional equipment was necessary for training and analysis purposes, recording of voice communications etc.

Running a typical direction center with its duplex AN/FSQ-7 computer, more than one hundred display consoles, the data communications equipment, the power supply systems etc. required substantial staffing as shown in table 4.2: Three-shift operation required at least 627 persons for a single DC peaking at about 1,000 people at maximum. Just the fact that at least 10 technicians were required for running the computer systems and its associated displays gives an impression of the challenge of running such an intricate installation under the extreme availability-requirements of the SAGE system as a whole.

4.4 AN/FSQ-7 overview

An AN/FSQ-7 installation consisted of a number of complex macroscopic subsystems:[44]

---

[43] See [SHULMAN 1959].

[44] These will be explained in more detail in the following chapters. A remarkably thorough description is also found in [Крысенко 1966], a Russian publication from 1966 covering AN/FSQ-7 and its surrounding systems in quite some detail.

Central computer system: The central computer system consisted in fact of a duplex AN/FSQ-7 installation where one of two identical computers had an active role while the other was typically in standby mode, tracking the activities of the active side or undergoing (preventive) maintenance. Each of the two computers consisted of several subunits: *program element, instruction control element, arithmetic element, I/O selection element,* and the *maintenance control element* (the maintenance consoles).

Main memory system: An AN/FSQ-7 computer contained two core-memory subsystems, one 4 k and one 64 k memory of 32 bits each, holding program instructions and data.

Maintenance consoles: These consoles allowed the operators and maintenance personnel to monitor and control the overall computer system operation. These extremely impressive consoles showed details of the inner state of the system like register contents, the operational status of input/output equipment etc.

Drum system: Like Whirlwind, AN/FSQ-7 featured several magnetic drum systems which were used for program and data storage and as buffers for asynchronous data transmissions.

Input system: This subsystem took care of incoming data from various sources like *LRI, GFI, XTL* from adjacent DCs and CCs, and *Manual Input (MI)*.

Output system: The output system was responsible for transmitting outgoing data to missile and interceptor bases and other SAGE centers in an asynchronous fashion with respect to the operation of the computer.

Display system: This subsystem consisted of the various display consoles and the associated equipment for their control.

Warning light system: Most display consoles featured warning lights which could be activated by the central computer requesting manual intervention by an operator. These lights were controlled by this subsystem.

Tape system: The computer also featured some IBM tape drives which could be used to write dumps for later analysis, load software distribution tapes, load simulated data for training or testing purposes, etc.

Card machines: IBM card readers were used for ad-hoc data input to the system containing information about the current weather situation or air traffic information.

Power and marginal checking system: Powering all these subsystems not only required a substantial amount of electrical power but also an intricate power distribution and marginal checking system based on the experiences gained with Whirlwind.

The interconnection of these subsystems in a single, so-called *simplex* AN/FSQ-7 computer is shown in figure 4.7. Due to the drum system effectively decoupling the central computer system from the main input/output devices, computer operation and

4.4 AN/FSQ-7 overview

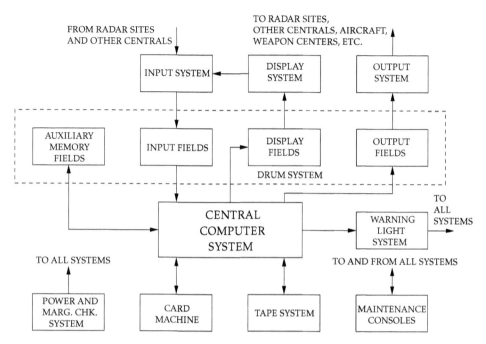

Figure 4.7: Schematic view of the AN/FSQ-7 Combat Direction Control (see [IBM CCS I][p. 3])

input/output operations could take place asynchronously. This made it possible to do without an intricate interrupt mechanism which would have complicated programming and operation of the computer considerably without adding to the necessary reliability of the overall installation. Figure 4.8 shows this decoupling of the various data flows to and from the central computer system in more detail. Even the interactive display consoles, which were connected to additional equipment like the *situation display generator*, the *digital display generator*, and the *manual input* system, were controlled by dedicated magnetic drums decoupling the activities of the computer and these subsystems, making completely asynchronous and parallel operation possible.

As explained earlier, a DC and CC was based on a duplex computer installation to achieve the extreme requirements regarding availability of the overall system. Thus a so-called *duplex switching facility* was necessary to allow operating two computers, one being the active computer essentially controlling all of the input/output equipment. Using this facility, a *switchover* between the two computers could be performed. Switchovers were either initiated in case of a computer failure or at regular intervals which allowed checking both machines without interrupting normal operations. The switchover caused the program running on the computer, the so-called *Direction Center Active (DCA)* in case of a DC or the *Combat Center Active (CCA)*, to be restarted. Each switchover required the execution of a *startover* to be performed on a computer. Basically the following modes of operation were possible:

Initiate: Restart of the computer after clearing all data in core memory. Information stored on magnetic drums was retained.

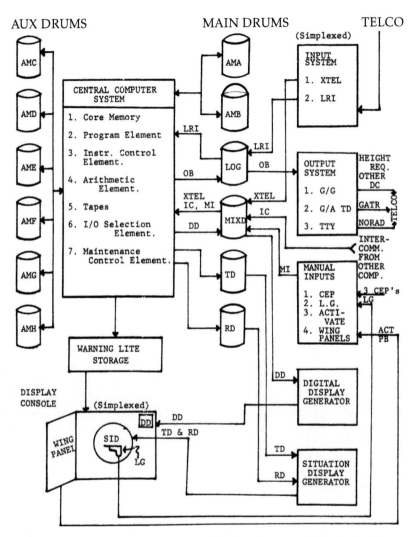

Figure 4.8: AN/FSQ-7 simplex system data flow (courtesy of Mike Loewen)

Brainwash: Similar to initiate, but data on magnetic drums were deleted, too.

Continue: The currently active computer was stopped and restarted without any loss of information about the current air situation etc.

Reestablish: This mode was similar to the continue-mode with the difference that data on the drums supplying the display system with information were checked for parity errors. Erroneous data were subsequently deleted from the drums during this process. As a result, the information about the current air situation could be up to three minutes old at the time of restart.

4.4 AN/FSQ-7 overview

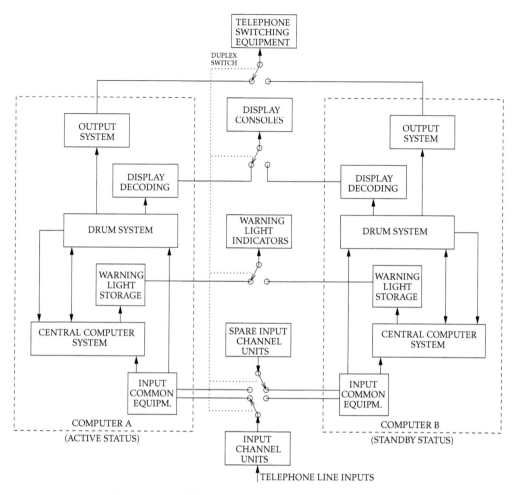

Figure 4.9: Structure of the Duplex Switching Facility (see [IBM CCS I][p. 5])

Maintenance personnel used the terms *duplex*, *simplex*, and *noplex* to describe the various states in which both computers, only one computer or none of the two computer systems was actively running the DCA or CCA.[45] Figure 4.9 shows the duplex switching equipment.

Figure 4.10 shows the layout of the many racks comprising one duplex AN/FSQ-7 installation which was housed on the second floor of an early blockhouse structure. Just right of the center, the maintenance console room can be seen. From there the two computer systems could be monitored and operated and start- and switchovers were executed. Input/output equipment was monitored from there as well, and two card punches in conjunction with two card readers could be used to load (test-)programs into the machines etc. The top and bottom half, rather symmetrically enclosing the

---

[45]See [FELSBERG 1969][p. 5].

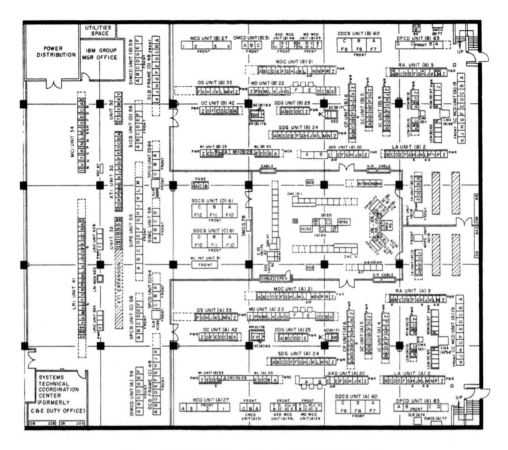

Figure 4.10: AN/FSQ-7 floorplan (see [IBM DRUM][p. 4])

maintenance console room, were occupied by the two AN/FSQ-7 computers while the left side of the machine room contained the electronics for GFI, LRI, and XTL as well as power supply and marginal checking systems. Table 4.3 lists the numbers and designations of the various units comprising the two computer systems and their associated input/output and power supply equipment.[46]

---

[46]Units 65, 66, and 67 are used in AN/FSQ-7 while these units were numbered 7, 8, and 9 for the AN/FSQ-8.

4.4 AN/FSQ-7 overview

Unit number	AN/FSQ-7 only	Description
1		Duplex maintenance console
2		Left arithmetic unit
3		Right arithmetic unit
4		Instruction control unit
5		Selection control unit
6		Program element
10 / 11 / 12		Core memory No. 2 (4 k)
13		Magnetic tape adapter
14 / 15 / 16 / 17		IBM type 728 magnetic tape drive
18		Tape power supply
19		Computer Marginal Checking and Distribution (MCD)
20		Auxiliary drum
21		Drum control
22		Main drum
23		Manual Data Input (MDI)
24		Situation Display Generator Element (SDGE)
25		Digital Display Generator Element (DDGE)
27		MI and display MCD
28		MI interconnecting
29		Drum MCD
30		Warning light storage unit
31		Output MCD
32		Crosstell input
33		Output storage
34	✓	GFI
41	✓	LRI
42		Output control
45		Duplex switching console
46		Auxiliary drum MCD
47		Simplex maintenance console
48		Display Circuit Breaker (CB)
51		IBM 713 card reader
52		IBM 718 line printer
53		IBM 723 card punch
55		Simplex input Power Distribution (PD)
56		Simplex CBs
58		Simplex Marginal Checking (MC)
59		Simplex common MCD
60		Duplex power supply
61		Simplex power supply
63		Duplex Power Control and Distribution (PCD)
64		Simplex PCD
		Continued on next page

		Continued from previous page
65 / 66 / 67		Core memory No. 1[47] (64 k)
91		Warning light interconnection unit
92		Test pattern generator
93	✓	LRI monitor control unit
94 / 95		Induction regulator
169 / 170	✓	Display console
177 / 178		Signal Data Patch Panel
250		Command Post Console
251 / 252		Photographic Recorder-Reproducer Element (PRRE)
352 / 353		IBM 020 Computer Entry Punch
600…617	✓	Mapper consoles
750 / 833		SD console
787	✓	SD console
935		Auxiliary display console

Table 4.3: Unit numbers and designations

---

[47] Originally this was a 4 k core memory system which was later replaced by a 64 k core stack.

5 Basic circuitry

The following sections focus on the basic circuitry used in AN/FSQ-7 – a necessary task for understanding and marveling the machine as such. While today's abundant *ultralarge scale* integrated circuits[1] easily contain hundreds of millions and even billions of active circuit elements in the form of transistors, a simplex AN/FSQ-7 had to make do with fewer than 30,000 vacuum tubes, resulting in a rather straightforward, yet quite powerful architecture for its time.

At first we will focus on the basic waveforms employed throughout the computer system, and a problem quite common in the era of AN/FSQ-7 and its solution will be detailed.

5.1 Introduction

Many of the circuits described in the following sections will look familiar to readers with some background in electronics, since most of the vacuum tubes used behave quite like *junction field-effect transistors*.[2]

Due to the sheer size of AN/FSQ-7, where cables were measured in meters, the circuit designers were confronted with the problem of parasitic oscillations in the computer circuits. Such oscillations must be avoided at all costs in a digital computer where even a single pulse transmits information. Figure 5.1 shows a very basic and nonsensical vacuum circuit on the left side, basically consisting of a single triode. This tube's plate is connected to the positive supply voltage through resistor R1 while its cathode and grid are connected directly to ground. In an ideal world this configuration would be stable and the tube would not conduct much current due to the grid being at ground potential.

Although every reasonable circuit designer would strive to keep the supply and control leads of the active elements as short as possible, the sheer size of AN/FSQ-7 made this impossible. This size was dictated not only by the fact that vacuum tubes, generating a great amount of heat, were employed – a further complication was the requirement for utmost serviceability which resulted in a rather sparse packaging. Elements were grouped into pluggable modules which could be removed and replaced easily by maintenance personnel. All this added to the unavoidable length of the wiring which resulted in the formation of parasitic resonant circuits. The right hand side of figure

---

[1] *ULSI* for short.
[2] *JFET* for short. In these devices current flowing through a *channel* between a source and a drain terminal can be controlled by a voltage – not a current as would be the case for a bipolar transistor. This control voltage *pinches* the channel, effectively controlling its conductivity.

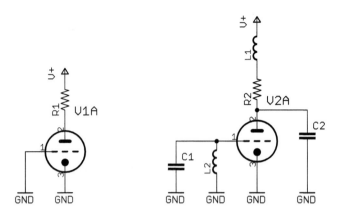

Figure 5.1: Parasitic inductances and capacitances in circuits

5.1 shows the same circuit as it results from these non-negligible lengths of interconnecting wires.

Between grid and ground, and plate and ground, two parasitic capacitances, C1 and C2, have to be taken into account, which are due to the construction of the vacuum tube itself. In addition to this, the long wires connecting the circuit to its supply and control voltages form unintended inductors L1 and L2. These two effects combined can and often will yield an unstable circuit which is subject to oscillations.

Mainly two techniques were employed in AN/FSQ-7 to suppress these oscillations: The first is to insert resistors into signal and power lines which are dimensioned in a way that the resonant circuit cannot oscillate due to the loss introduced by these resistors.[3] The second method is based on RC-combinations, combinations of a resistor and a capacitor which act as simple low-pass filter circuits in the supply lines. These RC-combinations have a crucial role since many switching circuits are connected to the same supply voltages. Without simple filters like these, spikes caused by the process of switching of one element could affect other elements connected to the same supply lines.

As can be seen in all of the following practical circuits, these RC-combinations often consisted of a 220 Ω resistor and a 10 nF capacitor. The *reactance* X_C of the capacitor is given by

$$X_C = \frac{1}{2\pi f C},$$

where f denotes the critical frequency at which decoupling in the circuit is required, while C represents the capacity. A typical value for f in AN/FSQ-7 is 1 MHz, where a 10 nF capacitor exhibits a reactance of about 15.9 Ω. A signal of that frequency or an impulse of corresponding duration will now "see" the capacitor as a rather low resistance compared with the 220 Ω resistor of the RC-combination. Since the capacitor

---

[3]It should be noted that these resistors are in series to the inductors formed by the long wires, while the unwanted capacitances C1 and C2 remain unaffected by these since these capacitances are caused by the tube's internal structures.

5.1 Introduction

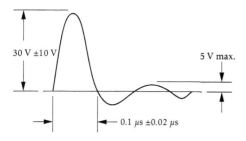

Figure 5.2: Standard pulse within AN/FSQ-7 (see [IBM BASIC CIRCUITS][p. 5])

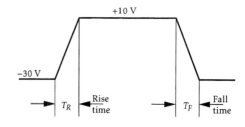

Figure 5.3: Standard levels within AN/FSQ-7 (see [IBM BASIC CIRCUITS][p. 5])

is connected to ground, it will effectively suppress that signal or pulse by providing a low-resistance ground path, thus decoupling the switching circuit from other circuits connected to the same power line.

Throughout AN/FSQ-7 two standard signals were used: A *standard pulse* and a *standard level* as shown in figures 5.2 and 5.3. The height of a standard pulse is about 30 V ± 10% while its width is 100 ns ±20 ns. These pulses were used to trigger various circuits like flip-flops etc. The other signal type is the standard level which is either at a potential of +10 V or at −30 V representing the two logical values of BOOLEAN algebra.

The basic symbols used in the schematics of AN/FSQ-7 for circuit elements and signals are quite different from what we are used to seeing today. Figure 5.4 shows the four different arrows used to denote standard pulses and levels and their respective non-standard counterparts.[4] These symbols are used to denote input as well as output lines of circuit elements.

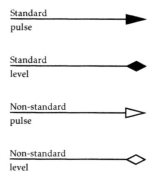

Figure 5.4: Signal symbols denoting pulses and levels (see [IBM BASIC CIRCUITS][p. 5])

The following sections now focus on the operation of the basic circuit elements used in AN/FSQ-7. In cases where different implementations for similar functions like a flip-flop were used, only one particular or at most few variation have been selected exemplarily for discussion.[5] Table 5.1 shows the basic circuit groups covered in the following sections. All basic building blocks can be grouped into logic and non-logic circuits which are then further detailed.

---

[4]Some circuits, like delay lines and some drivers, used pulses and levels with non-standard voltages.
[5]Detailed information on all circuits used can be found in [IBM BASIC CIRCUITS] and [IBM SPECIAL CIRCUITS].

Non-logic			Logic			
Power amplifiers	Control	Restorer	Coincidence	Conversion	Timing	Storage
Cathode followers	Thyratron relay drivers	DC level setters	Diode AND	Inverter	Delay unit	Flip-flop
Pulse amplifiers	Vacuum-tube relay driver		Diode OR	Pulse generator	Single shot	
Register drivers			Gate tube			

Table 5.1: Grouping of basic circuits (see [IBM BASIC CIRCUITS][p. 3])

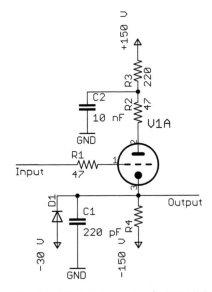

Figure 5.5: Basic schematic of a cathode follower (see [IBM BASIC CIRCUITS][p. 17])

5.2 Cathode follower

One of the most basic vacuum tube circuits of all is the *cathode follower*[6] which has been introduced in section 2.2. Figure 5.5 shows the schematic of such a cathode follower as used in AN/FSQ-7. Due to its high-impedance input and low-impedance output, the cathode follower acts as a non-inverting amplifier with approximately unity gain and accordingly belongs to the power amplifier group.

The deviations of this circuit from the basic cathode follower shown in figure 2.4 are mostly due to the necessity of suppressing parasitic oscillations as described before: First of all, the input signal is connected to the control grid of the triode through a series resistor counteracting the inductance introduced by the input wiring. The RC-combination consisting of R3 and C2 in the plate circuit of the tube effectively sup-

---

[6]See [IBM BASIC CIRCUITS][pp. 15 ff.].

5.3 Pulse amplifier

presses spikes which could otherwise be coupled to other circuits through the +150 V supply line. The cathode is connected to −150 V through R4 with a diode clamping the output signal to −30 V at minimum. The actual value of R4 varies for different types of cathode followers. The AN/FSQ-7 used eight different variants with typical values of R4 between 17.1 kΩ and 40.1 kΩ or no R4 at all. The capacitor C1 decreases the fall time T_F of the output signal, as it provides a low-resistance ground path for high frequency signals. T_F is also affected by the size of R4 – the smaller this resistor, the shorter the fall time.

The abstract symbol for such a cathode follower is shown in figure 5.6. Input and output as denoted here are standard levels, and no particular variation of the circuit has been specified. A specific module will be denoted by a small capital letter B, C, D, E, F, G, H, or J to the left of "CF".

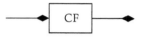

Figure 5.6: Symbol of a cathode follower

5.3 Pulse amplifier

The second element of the power amplifier group is the *pulse amplifier*.[7] This circuit amplifies a standard pulse so that a higher load can be driven. In modern parlance a pulse amplifier increases the *fan-out* of a signal. AN/FSQ-7 used three different models of this amplifier, denoted by A, B, and C. Type A can drive a constant light load of two to eight load units,[8] type B is capable of driving a constant heavy load of five to eleven load units, while type C was used in case of varying light loads between zero and three units. The abstract symbol of a pulse amplifier is shown in figure 5.7. As its name implies, this circuit element requires and yields standard pulses instead of levels.

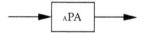

Figure 5.7: Symbol of a type A pulse amplifier (type B and C would be denoted by "BPA" and "CPA", respectively)

Figure 5.8 shows a model A pulse amplifier for constant light load. At its heart is the pentode V1 which has its *suppressor grid* tied to a slightly positive voltage through the RC-combination R6 and C4, while its *screen grid* is at an even higher potential due to R5 and C3. The output circuit of this pulse amplifier consists of the transformer T0 which inverts the polarity of the output pulse and serves as an impedance matching element for the elements connected to its secondary. Its turn ratio is 4:1 resulting in an impedance ratio of 16:1.

The control grid of V1 is biased at −15 V through R3 and R2 (C2 smoothes this bias voltage). The input signal is decoupled from this biased grid circuit through capacitor C1. An incoming standard pulse counteracts this negative grid bias so that V1 begins to conduct. The pulse current flowing through the pentode generates an output-pulse at the secondary of T0. The diode D1 creates a low-resistance discharge path for the input coupling capacitor C1. The resistor R1 terminates the input line to avoid reflection of pulses back to the generating circuit.

---

[7]See [IBM BASIC CIRCUITS][pp. 25 ff.].

[8]A load unit represents the resistive and capacitive loading of a typical circuit element used in AN/FSQ-7.

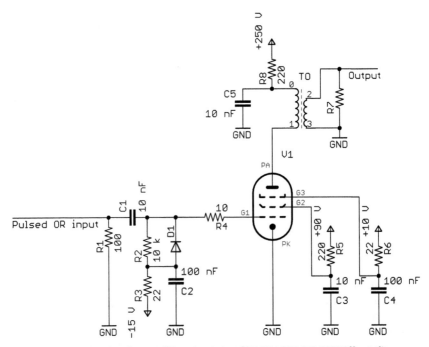

Figure 5.8: Type A pulse amplifier circuit (see [IBM BASIC CIRCUITS][p. 26])

The type B pulse amplifier is similar to this type A variant. The only difference concerns the suppressor grid which is tied to the plate by a 10 Ω resistor instead of being tied to the +90 V potential.[9] The type C circuit is based on type B with a modified input network.

The input circuit shown in figure 5.8 is a so-called *pulsed OR input* while a slightly modified input network in which D1 is omitted and R2 is replaced by a resistor in series with a coil was called *direct input*.[10]

5.4 Register driver

Figure 5.9: Symbol of a type A register driver (type B would be denoted by "ʙRD")

The last element of the power amplifier group is the *register driver*,[11] the symbol of which is shown in figure 5.9. Basically this device consists of several pulse amplifiers driven by a common input signal. Types A and B differ only marginally with respect to their input network. The detailed schematic of a type A register driver is shown in figure 5.10.

---

[9] The pentode is operating as a tetrode.
[10] See [IBM BASIC CIRCUITS][p. 26].
[11] See [IBM BASIC CIRCUITS][pp. 31 ff.].

5.4 Register driver

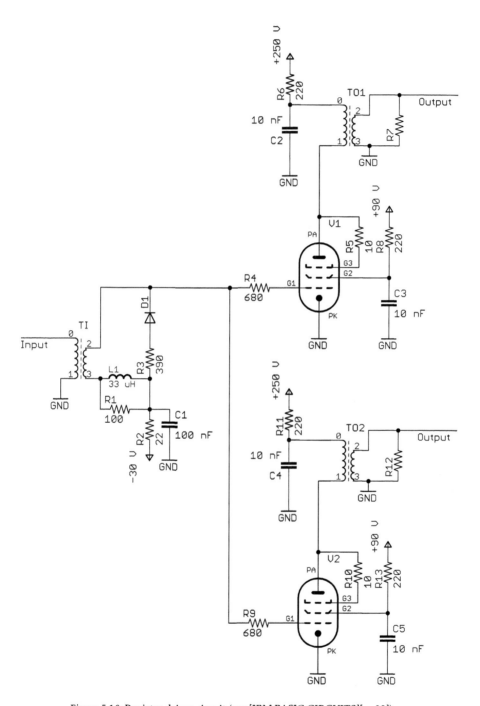

Figure 5.10: Register driver circuit (see [IBM BASIC CIRCUITS][p. 32])

The only difference with respect to a pulse amplifier is the input stage which decouples the input signal from the biased grids of the output stages by means of the transformer TI. The grids are biased at -30 V through R1, R2, and the secondary winding of TI. A standard pulse applied to the primary side of TI results in a positive pulse at its secondary, followed by a rather large negative undershoot due to resonance effects. This negative undershoot is limited by D1 and R3. The coil L1 in parallel to R1 helps suppressing the generation of a reflected signal at the primary of TI.

5.5   Relay drivers

The next section of the non-logic group contains two relay driver circuits, the *thyratron relay driver*[12] and the *vacuum tube relay driver*.[13] First of all the principle of operation of a thyratron should be explained:

A *thyratron* is essentially a gas-filled tube which can best be compared with today's *thyristors*.[14] The thyratrons used as relay drivers in AN/FSQ-7 were Xenon filled *tetrodes*, i.e. tubes with two grids, one control grid and one *shield grid*. In contrast to a vacuum tetrode, a positive potential at the control grid can be used to *trigger* the thyratron in such a way that a conducting path of ions is generated between plate and cathode of the thyratron. As a precondition, the potential difference between plate and cathode must match or exceed the *ionization potential* of the tube which is determined by various factors such as the tube's geometry, the filling gas, and its pressure. When a thyratron is conducting, the control grid becomes ineffective due to a cloud of positively charged ions around it, so the only way to switch off a thyratron is to lower the plate-cathode potential difference to a level much lower than the ionization potential. This level is called the *extinction potential* of the tube.

Figure 5.11 shows the circuit of a model A thyratron relay driver:[15] With the circuit breaker closed, the potential between plate and cathode is well above the necessary ionization potential. Since the control grid is negatively biased through R2 and R3, the tube will not conduct. A positive incoming pulse will raise the grid potential through the decoupling capacitor C3, thus triggering the thyratron. As a result of this, the current flowing through V1, R5 and the circuit breaker will activate the relay K1.[16] To reset the relay driver, the circuit breaker must be opened either manually or automatically, so that the ionization potential is no longer maintained between plate and cathode.[17]

---

[12] See [IBM BASIC CIRCUITS][pp. 35 ff.].

[13] See [IBM BASIC CIRCUITS][pp. 39 ff.].

[14] In fact, the name thyristor is the combination of "thyratron" and "transistor", showing the heritage of this circuit element.

[15] The model B thyratron relay driver uses the shield grid of the thyratron to implement a simple AND-gate, so that relay activation can only take place when this second input line is at positive potential. Otherwise the driver circuit is blocked. There was also a model D thyratron driver which is described in [IBM SPECIAL CIRCUITS][pp. 161 ff.].

[16] This circuit is thus called being *plate loaded* since the load is part of the plate circuit. A variation of this thyratron driver, and the type B driver as well, uses a cathode loaded configuration where the relay is located between cathode and ground.

[17] The RC-combination C4 and R6 forms an arc-suppressor network protecting the circuit breaker.

5.5 Relay drivers

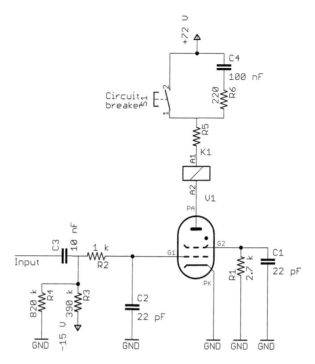

Figure 5.11: Basic form of a thyratron relay driver (see [IBM BASIC CIRCUITS][p. 36])

Figure 5.12 shows the abstract circuit symbol for such a type A thyratron relay driver.[18] Due to the high amount of current a thyratron can switch, this type of driver was mainly used to drive multiple relays, print and punch magnets or *wire contact relays*.[19]

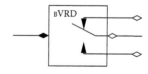

Figure 5.12: Symbol of a Type A thyratron relay driver

The symbol of a vacuum tube relay driver is shown in figure 5.13. Since the triodes used in this circuit (see figure 5.14) are not capable of driving currents as high as those of the thyratron circuit, typically only a single low-energy relay can be controlled by these drivers. Accordingly, the symbol incorporates the relay contacts.

The circuit is quite straightforward: The triode V1A is controlled by a standard level applied to its grid through resistor R1. When the grid becomes sufficiently positive with respect to the cathode, i. e. at about +10 V, the tube begins to conduct and the current flowing from cathode to plate energizes relay K1 through R2 and R3. R3 and C2 form

Figure 5.13: Symbol of a vacuum tube relay driver

---

[18]The type B symbol features an additional standard level input. The relay will be activated only if both inputs are at +10 V level.

[19]Wire contact relays were developed and used by IBM for applications requiring a high-speed relay that could be operated at 40 V DC, see [IBM RELAYS][pp. 33 ff.].

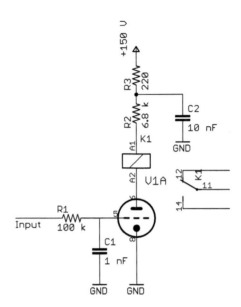

Figure 5.14: Basic form of a vacuum tube relay driver (see [IBM BASIC CIRCUITS][p. 39])

an RC-combination to make sure that no spikes from switching the relay will enter the supply rail. An additional means toward this is C1 which slows down the rise time of the grid voltage since the transition from −30 V to +10 V of the input signal must change the charge stored in this capacitor. This results in a slower rise of the current flowing through the relay. The same effect takes place when the input signal falls.

5.6 Level setter

The so-called *level setter*[20] is a member of the restorer group of table 5.1. It employs a cathode follower as its output stage while the input circuit consists of another cathode follower driving a so-called *grounded grid amplifier*. The purpose of the level setter is to restore proper signal levels of an input signal. There are two types of this device: Model A accepts an input signal with a high-level between +12 V and 0 V and a low-level between −30 V and −8 V, while model B requires an input signal with a high-level between +12 V and +6 V and a low-level between −30 V and −11 V.

Figure 5.15: Symbol of a type A level setter (type B would be denoted by "ʙLA")

Figure 5.15 shows the abstract symbol of a model A level setter as used in the AN/FSQ-7 schematics. Figure 5.16 shows the detailed schematic of a model A level setter.

The input stage cathode follower is dimensioned so that a negative input voltage connected through R1 to the grid results in a negative output voltage at the cathode of V1A. The grid of the second triode V1B is connected to a voltage divider consisting of resistors

---

[20]See [IBM BASIC CIRCUITS][pp. 19 ff.].

5.6 Level setter

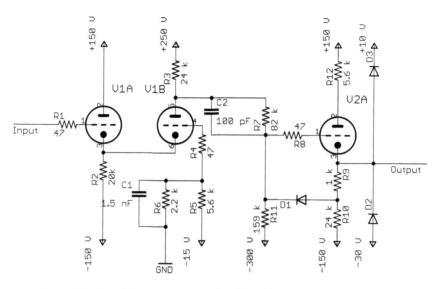

Figure 5.16: Type A level setter circuit (see [IBM BASIC CIRCUITS][p. 22])

R5 and R6, which is connected between the −15 V supply and ground. By means of this voltage divider, the grid is at a slightly negative fixed potential with respect to ground.

When the cathode potential of V1A becomes negative due to a negative input level, the cathode potential of V1B follows. As a result of this, the grid of V1B becomes positive with respect to its cathode and the tube begins to conduct. Thus the plate potential of V1B drops due to the current flowing through R3 and R2. Accordingly, the grid potential of the output stage triode V2A, which is connected to the voltage divider consisting of R7 and R11, becomes negative and the tube stops conducting and its cathode potential will become negative with respect to ground. Due to the clamping diode D2, the output of this last stage is limited to −30 V, the desired standard level potential.

In case of a positive input signal, things reverse: V1A conducts, so its cathode potential will become positive, making the cathode of V1B positive, too. The grid potential of V2A will then be negative with respect to its cathode and V2B does not conduct. As a result, the grid of V2A becomes positive with respect to its cathode and the triode will start to conduct. Its output will rise to a rather positive voltage which is clamped by D3 to +10 V.

The type B level setter differs from type A in various respects: First of all, the cathode of V1B is no longer connected to the cathode of V1A directly but through a voltage divider which replaces R2, thus effectively changing the input signal thresholds. Second, all positive supply voltages are decoupled by RC-combinations consisting of a 220 Ω resistor and a 10 nF capacitor to reduce injection of switching noise into the supply lines.

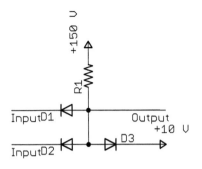

Figure 5.17: Basic two-input AND circuit (see [IBM BASIC CIRCUITS][p. 42])

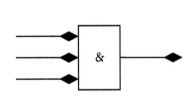

Figure 5.18: Symbol of an AND circuit (see [IBM BASIC CIRCUITS][p. 41])

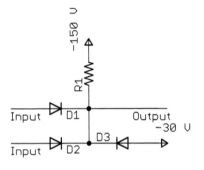

Figure 5.19: Basic two-input OR circuit (see [IBM BASIC CIRCUITS][p. 43])

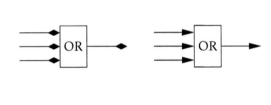

Figure 5.20: Symbol of an OR circuit (see [IBM BASIC CIRCUITS][p. 43])

5.7 Diode AND and OR circuits

Gates implementing logical AND and OR operations belong to the coincidence group of table 5.1 and are based on simple diode networks in AN/FSQ-7. Figures 5.17 and 5.18 show the schematic of a two-input AND gate as well as the abstract symbol for an AND circuit with three input lines.[21] The principle of operation is simple: When at least one of the two inputs is at −30 V potential, it will tie the output down to this voltage. Only if both inputs are at +10 V, the output will go high.

In principle, the diode D3 would be unnecessary as it just acts as a third input to the AND gate that is always tied to +10 V. Its purpose is to protect following vacuum tube circuits from excessive grid levels[22] in case of inputs to the AND gate being driven way beyond +10 V due to a component failure in an earlier stage. In this case, D3 clamps the output signal to +10 V.

---

[21] More inputs were, of course, possible by using additional input diodes.

[22] When the control grid of a vacuum tube becomes too positive with respect to the cathode, it will act as a parasitic anode and start collecting electrons from the cathode. In addition to excessive overall current flowing through the tube in such a case, this will in turn generate a substantial grid current which will heat the grid. These two effects can destroy a vacuum tube and must therefore be avoided.

5.8 Gate tube circuit

OR gates were based on the same scheme but with reversed diodes and reversed bias voltage as shown in figure 5.19. Due to D3, the output voltage is clamped at a maximum negative voltage of −30 V. If both inputs are at −30 V a current will flow through D1 and D2 yielding −30 V at the output. If, e. g. diode D1 is at +10 V potential at its anode, a current will flow through it, effectively raising the output potential to +10 V, too. If the second input is still at −30 V, D2 will stop conducting since its cathode is now at a higher potential than its anode, thus a single input – more are possible by adding input diodes to this circuit – can force the output to +10 V potential.

The OR circuit exhibits a much faster signal rise time than the AND circuit due to the latter's higher load capacitances. Accordingly, the OR circuit may be used with standard levels as well as standard pulses as inputs while the AND circuit works only with standard levels. Figure 5.20 shows both variations of the abstract symbol for a three-input OR circuit, the (standard level) diode OR and the *pulsed OR*.

5.8 Gate tube circuit

The *gate tube circuit*[23] is part of the logical elements of table 5.1. It implements a coincidence function (quite like an AND) where a standard level input can be used to block or unblock a second input which accepts standard pulses; something that would not be possible with a diode AND due to its slow response time. These two states were called *conditioned* and *non-conditioned* respectively. The graphical symbol for a model A gate tube circuit is shown in figure 5.21 – the control input is on the top of the element with the pulse input and output on the left and right hand side.[24]

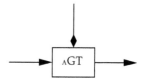

Figure 5.21: Symbol of a model A gate tube

The rather straightforward schematic of this circuit is shown in figure 5.22: The active component is a pentode, having its control grid connected to a direct pulse input with an input network similar to those described before, while its suppressor grid is connected to the level input controlling the conditioned and non-conditioned state. The screen grid is tied to +90 V by R2 and C1. The output section is similar to that of a pulse amplifier using a transformer for decoupling and impedance matching. Note resistors R6 and R7 for terminating the pulse input and output lines, respectively.

The use of coil L1 is interesting: During the quiescent state of the circuit, i. e. without an input pulse, the control grid is negatively biased through R4 and L1; R5 and C2 act as a simple low-pass filter. For an incoming pulse L1 exhibits a very high resistance due to its inductance. After the pulse, L1 provides a low-resistance path for biasing the grid and for discharging the coupling capacitor C3, thus reducing recovery time of the overall circuit.

---

[23] See [IBM BASIC CIRCUITS][pp. 45 ff.].
[24] No evidence for a model B gate tube circuit could be found.

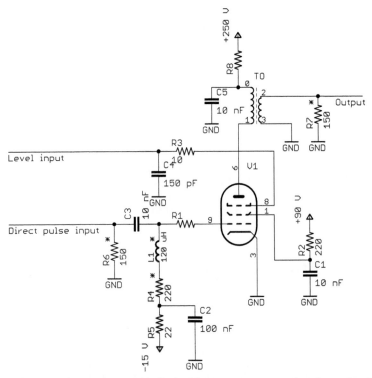

Figure 5.22: Model A gate tube circuit, parts marked with a star may vary as they determine the switching behavior (see [IBM BASIC CIRCUITS][p. 46])

5.9 DC inverter

Figure 5.24: Symbol of a model A DC inverter

The purpose of the so-called *DC inverter*[25] is to accept a level input and output an inverted signal with standard level. The symbol of such an inverter[26] is shown in figure 5.24. The circuit shown in figure 5.23 consists of two sub-circuits: The input signal controls the grid of a heavily over-driven amplifier[27] which performs the actual inversion of the signal. The tube V1A will be blocked for a negative input signal and conduct for a positive one.[28] When the tube conducts, the voltage at its plate will be near ground, otherwise it will be at a quite positive potential due to R3.[29]

The signal at the plate of V1A controls the cathode follower consisting of V1B. The grid of this tube is negatively biased through R7 and R6. This bias can be overridden by a

---

[25] See [IBM BASIC CIRCUITS][pp. 49 ff.].
[26] There was only a model A DC inverter.
[27] This means that the amplifier generates an output signal that is always saturated or near ground.
[28] Blocking will occur for input voltages at or below −8 V while any level between 0 V and +12 V will make the tube conducting heavily. So this circuit also serves as an inverting level restorer, since no standard level input signal is necessary.
[29] R4 and C2 serve as a decoupling RC-combination.

5.10 Flip-flops

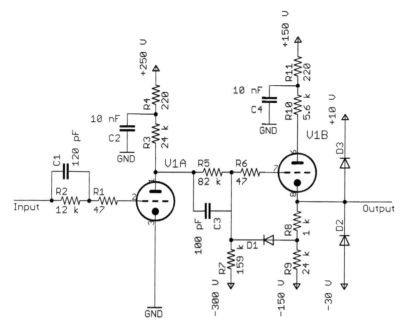

Figure 5.23: Model A DC inverter (see [IBM BASIC CIRCUITS][p. 50])

sufficiently positive potential from the plate of V1A.[30] The diodes D2 and D3 clamp the output signal to −30 V and +10 V levels respectively completing the output stage.

5.10 Flip-flops

One of the most basic, yet central elements in a digital computer is of course the flip-flop. The AN/FSQ-7 featured three basic flip-flop models[31] as shown in table 5.2 which also contains the abstract circuit symbols used to denote these building blocks. Two of these flip-flops are discussed in the following, the model B low-speed[32] flip-flop and its model A high-speed counterpart.

The circuit of the low-speed flip-flop is rather simple as it closely resembles that shown in figure 2.6, section 2.2: Two inverting amplifiers, consisting of the tubes V1A and V1B, are connected in a crosswise fashion with the plate of one tube driving the other's control grid and vice versa. Since the circuit is symmetric in its nature and must exhibit two stable states, the corresponding resistors R7 and R8, R5 and R6, R3 and R12, R2 and R11 must have corresponding values.[33]

---

[30] Note that the plate of V1A is connected to the +250 V rail through R3 and R4 while that of V1B is connected to +150 V potential through R10 and R11.
[31] See [IBM BASIC CIRCUITS][pp. 51 ff.].
[32] The low-speed flip-flop is capable of operation at up to 500 kHz and up to 400 kHz with inverted inputs.
[33] The same holds true for the RC-combinations at the −150 V rail, of course.

Table 5.2: Flip-flop types and their respective characteristics (see [IBM BASIC CIRCUITS][p. 51])

Model	Logic block symbol		Characteristics	
	Pulse input	Level input	Speed	Drive capability
A	AFF		High-speed	Can drive load directly
B	BFF	BFF	Low speed	Cannot drive load directly
C	CFF	CFF	Even lower speed	Can drive load directly

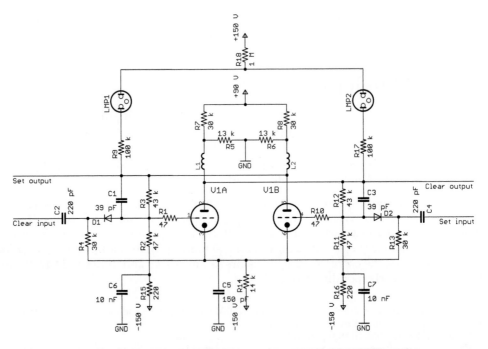

Figure 5.25: Model B low speed flip-flop (see [IBM BASIC CIRCUITS][p. 54])

5.10 Flip-flops

If the flip-flop is in the set-state, V1A is conducting. Accordingly, the potential at its plate is negative, so that tube V1B is cut off due to the negative potential at its control grid. Thus, the left output is at a positive voltage while the right output is negative. The upper and lower levels for these output lines, +10 V and −30 V respectively, are determined by two voltage dividers for each output. The output voltage for the left output is determined by the voltage divider consisting of R6 and R8 in the plate circuit of V1B and a second divider consisting of the resistors R3 and R2.[34] The same holds true for the right signal output, where the voltage dividers are R7, R5 and R12, R11.

A negative transition of the left input will cause a negative pulse at the cathode of D1, due to the differentiation performed by C1 and R1. Only negative pulses can pass diode D1 and will cut off V1A which will in turn change the voltage at its plate to a positive value turning on V2B. Now the right output is at +30 V while the left is at −10 V. To speed up transitions in either direction, resistors R3 and R12 are paralleled by capacitors C1 and C3.[35] In addition to that, two coils, L1 and L2, are wired in series to the plates of V1A and V1B, acting as so-called *peaking coils*.

The input circuits shown here, C2, R3, D1, and C3, 13, D2 are used for level inputs. If this model B flip-flop is to be used with standard pulses as input, each input is connected to the secondary of a pulse-transformer. This secondary winding is shunted by an additional diode against ground so that positive levels won't reach the differentiating capacitor of the flip-flop. The primary side of each of the two pulse-transformers can be driven by a number of pulse inputs each decoupled by a dedicated diode which is oriented in such a way that only positive pulses can reach the transformer.

The reason that this type B flip-flop cannot drive external loads directly is due to the way the output signals are generated: Since there is no active clamping by diodes, the output voltage will deviate more or less from its nominal values of +10 V and −30 V depending on the load. Thus, this flip-flop can only drive DC inverters or level setters which are then connected to additional circuitry. The difference between this and the model C flip-flop is that the latter uses a slightly different scheme for generating the two output signals including two clamping diodes to ensure a proper output level even under load.

Figure 5.26 shows the detailed schematic of a model A high-speed flip-flop capable of operation at up to 2 MHz pulse repetition rate.[36] The main difference to the circuit discussed above is the addition of two cathode follower stages comprised of the tubes V1A and V2A which serve a double purpose: First, they allow driving a load directly at the two output lines. Second, they decouple the control grid of V2B from the plate of V1B and vice versa thus speeding up the overall operation of the flip-flop. Since the cathode follower V1A amplifies the output signal at the plate of V1B, both triodes are contained in the same glass envelope to minimize wire length and thus stray capacitances in the circuit further speeding up its operation. Another difference to the model B flip-flop is the higher plate voltage of +150 V.

---

[34] The RC-filter consisting of R15 and C6 can be neglected due to the small value of R15. The coil L2 can be neglected, too, due to its small DC resistance.

[35] Compare figure 2.5 in section 2.2.

[36] It is remarkable that this circuit is nearly identical to the high-speed flip-flop used in Whirlwind, see figure 3.8, section 3.

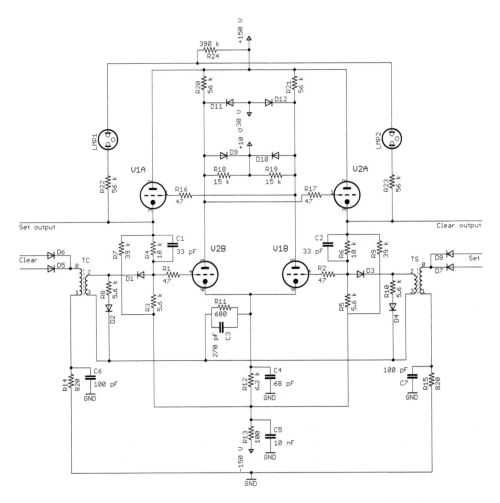

Figure 5.26: Model A high-speed flip-flop (see [IBM BASIC CIRCUITS][p. 58])

5.11 Single-shots

A *single-shot*[37] as has been used in AN/FSQ-7 is triggered by a standard pulse and generates a (non-)standard level output signal with a specified duration t. This output signal can either be at a negative level during the quiescent state of the single-shot and raise to a positive level or vice versa. Three different models, B, C, and D, were used as standard components. Figure 5.28 shows the abstract circuit symbol of such a single-shot. As before, the particular model is denoted by a small capital letter.

Figure 5.27 shows the different behavior of the model B, C, and D single-shots. All three are triggered by a standard pulse shown at the very top of the figure. While model B and C generate a standard level output signal, type D yields an output signal

---

[37] See [IBM BASIC CIRCUITS][pp. 61 ff.] and b [SAYRE 1949][pp. 166 ff.].

5.11 Single-shots

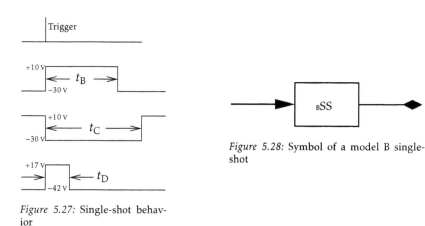

Figure 5.27: Single-shot behavior

Figure 5.28: Symbol of a model B single-shot

swinging between −42 V and +17 V. Since single-shots exhibit only one stable state they are also known as *monostable multivibrators, one-shot multivibrators*,[38] or *gating multivibrators*.

Figure 5.29 shows the detailed schematic of a model B single-shot, yielding a positive amplitude with a duration of t_B at the standard level output when triggered. The heart of the circuit consists of the tubes V1A and V1B. V2A is just an output stage based on a basic cathode follower. While the single-shot is in its stable (quiescent) state, V1B is conducting since its grid, being at ground potential, is more positive than its cathode. Accordingly, the junction between R8 and R9 is at −30 V due to the clamping diode D5 so that diode D3 conducts. This holds V1A in its cutoff state as its grid is now more negative than its cathode which is clamped to −15 V by means of D2, so the single-shot remains in its current stable state.

An incoming standard pulse trigger is inverted by transformer TI. A negative pulse reaches the plate of V1A through D1 (the coil L1 is a high resistance path for such a pulse while it exhibits low DC-resistance). From there it reaches the grid of V1B through C3 thus lowering the current through this tube, which results in a rise of its plate potential. This causes the negative potential at the grid of V1A to rise accordingly, and V1A starts to conduct.

This, in turn, decreases the plate potential of V1A which increases the effect of the initial negative pulse as it couples through C3 to the grid of V1B. Eventually, V1B will be cut off completely. The high plate potential at this tube is clamped by diode D4 to +10 V and amplified by the cathode follower V2A which yields the output signal which is now positive.

Now that V1B is cut off, the coupling capacitor C3 is discharged through R6, so the time-constant of the RC-network determines the length of the standard level output signal generated by this circuit. As a result, the potential at the grid of V1B rises to the point where the tube starts conducting again. Thus the potential at the junction of R8 and R9 falls to −30 V since it is clamped by diode D5 which causes the output of the

---

[38] Often abbreviated as *one-shot*.

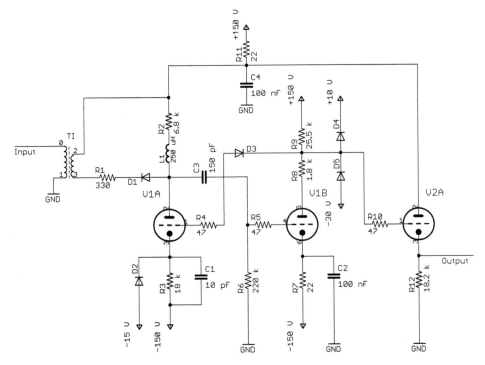

Figure 5.29: Model B single-shot (see [IBM BASIC CIRCUITS][p. 64])

cathode follower V2A to go low, too. In addition to this, V1A is cut off again through D3 and R4.

The resistors R4, R5, and R10 suppress parasitic oscillations which could be caused by the inherent grid-capacitances of the tubes. R11, C4 and R7, C2 form basic decoupling networks for the positive and negative supply lines while C1 creates a low reactance path for the cathode of V1A, reducing its switching time.

The input circuit shown in figure 5.29 – consisting of TI, R1 and D1 – is the so-called *low-speed input* as it is used for output signal-widths $4~\mu s \leq t_B \leq 10^5~\mu s$ with $68~pF \leq C3 \leq 500~nF$ and $220~k\Omega \leq R6 \leq 700~k\Omega$. In cases where a much shorter output-pulse width $1~\mu s \leq t_B \leq 4\mu s$ is required, this simple input circuit does not suffice. An active input circuit is necessary then:[39] A fourth triode is added which is connected with its plate to the plate of V1A, while its cathode is connected to +10 V. The grid of this tube is connected to the secondary of an input transformer like TI. This transformer does not invert the trigger pulse so that this tube conducts heavily during an input pulse, pulling down the plate of V1A faster than with the simple input circuit shown in figure 5.29.

The model C single-shot is based on model B with an additional inverter for the output signal, while model D features an output stage consisting of two paralleled cathode followers capable of driving high loads.

---

[39] TI, R1 and D1 will be removed accordingly.

5.12 Pulse generators

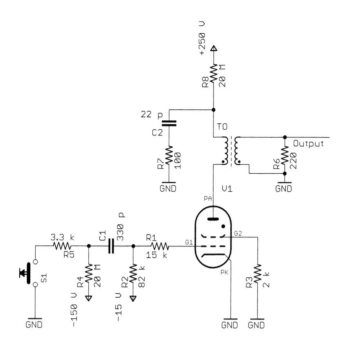

Figure 5.31: Model A pulse generator (see [IBM BASIC CIRCUITS][p. 72])

5.12 Pulse generators

While a single-shot converts a standard pulse into a standard level signal, a *pulse generator*[40] does more or less the opposite: An incoming pulse or a standard level signal

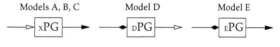

Figure 5.30: Pulse generator symbols

which may be generated by a manually operated button or a cam-controlled switch, is converted into an output pulse. All in all, five different pulse generator models, denoted by letters A through E, were used. The symbols for these devices are shown in figure 5.30.

The models A, B, and C feature a non-standard pulse input and are based on a thyratron – their circuit, shown in 5.31, is rather similar to that of a thyratron relay driver: Pressing the switch S1 causes a positive going pulse through C1 which ionizes the thyratron, lowering its plate potential. This, in turn, generates an output pulse on the secondary side of T0. The main difference to the thyratron relay driver is the RC-combination R7, C2: During the quiescent state, C2 is at +250 V through R8. When the thyratron ionizes, a rather large current flows through V1. This discharges C2 until the potential difference between anode and cathode of the thyratron is below the ionization threshold thus extinguishing the thyratron automatically.

---

[40] See [IBM BASIC CIRCUITS][pp. 71 ff.].

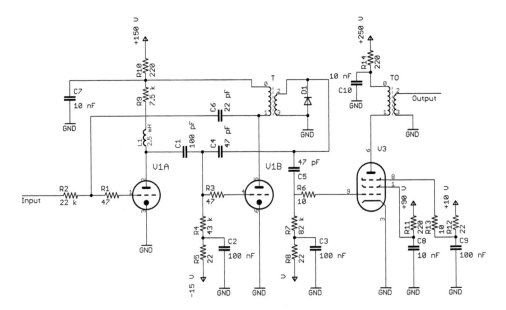

Figure 5.32: Model E pulse generator (see [IBM BASIC CIRCUITS][p. 75])

In some cases, due to a bouncing switch S1 etc., this simple reset circuit is not sufficient since multiple output pulses might be generated through a single activation of S1. In such cases a *normally closed* (NC for short) switch is connected in series with R8. This switch is then built as part of S1 in such a way that it opens just before S1 closes, so only the energy stored in C2 is available to generate a single output pulse.

The models A and B differ only with respect to R4 which is 200 kΩ for a type B pulse generator. Model C does not feature the switch input circuit consisting of S1, R5, and R4, instead the input pulse is coupled directly to capacitor C1.

The model D and E pulse generators are based on the principle of the so-called *blocking oscillator*.[41] Figure 5.32 shows the detailed circuit diagram of model E which expects a standard level input of +10 V during its quiescent state and generates an output pulse when this level drops to −30 V. Its output stage is straightforward and consists of a pentode driving a pulse transformer T0. The interaction of the two triodes V1A and V1B deserves more attention: With a +10 V input level, V1A is conducting since its grid is well above ground potential, while V1B is cut off since its grid is tied to −15 V by the resistors R3, R4, and R5.

When the input signal changes to −30 V, V1A is cut off and its plate potential rises to +150 V. This signal is differentiated by C1 causing a positive pulse at the grid of V1B which begins to conduct. This causes a current to flow through the primary of transformer T which results in a positive going pulse at its secondary. This triggers the output pentode through C5 and causes the grid of V1B to rise even higher. This increase in grid potential causes a grid current to flow thus charging both coupling capacitors

---

[41] See [MacNichol et al. 1949].

5.13 Delay lines and delay line drivers

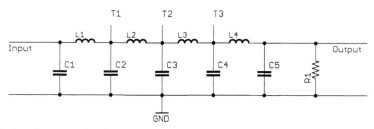

Figure 5.33: Basic schematic of a discrete-constant delay line (see [IBM BASIC CIRCUITS][p. 77] and [HUGHES 1949][p. 730])

Figure 5.34: Typical delay line setup (see [IBM BASIC CIRCUITS][p. 78])

C1 and C4. Since the transformer T actually performs a differentiation operation, the pulse generated at its secondary will fall back to ground potential causing the right hand side of C4 to be grounded. This adds to the negative grid bias of V1B, effectively blocking the tube. As soon as C1 and C4 have discharged through R4 and R5, the pulse generator is ready for the next falling edge of its input signal.

5.13 Delay lines and delay line drivers

So-called *delay lines* had been in use since World War II to delay complex signals like those delivered from radar devices for several μs.[42] Basically, a delay line consists of a series of LC-filters, combinations of coils and capacitors as shown in figure 5.33.[43] An arrangement which is called a *discrete-constant* line.[44]

An incoming pulse charges C1, thus raising the potential across C1. As a result, a current starts to flow through L1 which in turn charges C2 and so on. The resistor R1 terminates the output of this discrete-constant delay line so that a pulse will not be reflected back to the input again. A typical so-called *delay unit* exhibits a delay of 500 ns while the taps T1, T2, and T3 allow shorter delays, too. Figure 5.34 shows a typical delay unit arrangement where three 500 ns delay units are connected in series giving an overall delay of 1.5 μs.

While single delay units can be driven by a standard pulse amplifier as shown in section 5.3, chains of delay lines require a special driver circuit, the so-called *delay line*

---

[42] Some early bit-serial digital computers like the Bull Gamma-3 used such delay lines as storage devices.

[43] See [HUGHES 1949] and [MACNICHOL et al. 1949][pp. 238 ff.] for a thorough treatment of delay lines in general.

[44] In some applications, although not in the AN/FSQ-7 itself as it seems, special delay cables instead of such discrete arrangements of circuit elements were also used. These combined the inductive and capacitive parts shown in figure 5.33 and are called *distributed-constant* delay lines accordingly.

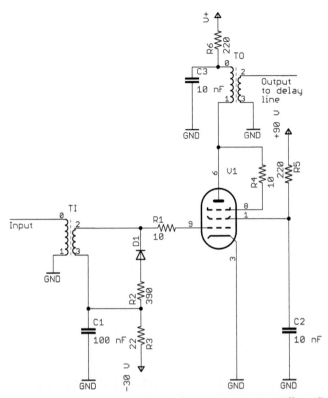

Figure 5.35: Delay line driver circuit (see [IBM BASIC CIRCUITS][p. 78])

driver. Figure 5.35 shows the schematic of such a driver: The control grid of the pentode V1 is negatively biased at -30 V through R3, the secondary of TI, and the parasitic suppressor R1, making sure that V1 is cut off initially. An incoming pulse at the primary of TI will cause a positive pulse on its secondary which in turn causes V1 to conduct. The current flowing through R6, the primary side of T0, and V1 causes the generation of a high-power output pulse on the secondary of T0 which is connected to the input of the delay unit chain to be driven. Since the suppressor grid of V1 is connected to the anode, the pentode is actually operating as a tetrode allowing a higher current to flow through V1.

5.14 Special circuits

Although the basic circuits described above were the main building blocks of an AN/FSQ-7 installation, there was a plethora of so-called *special circuits*[45] necessary for various purposes: Core memory row-, column-, and inhibit-drivers, sense amplifiers amplifying the tiny currents induced in the sense wires, core shift circuits

---

[45] See [IBM SPECIAL CIRCUITS].

to convert parallel data into serial data-streams suitable for data transmission over telephone lines, pulse shapers for amplifying the amplitude and reducing the width of pulses, binary decoders[46] to generate the deflection voltages for the various displays, deflection amplifiers, drivers, and vector generators for displays, amplifiers for area discriminators and light guns, crystal oscillators, data transmission and receiving circuits, drum reading and writing amplifiers, SCHMITT trigger circuits, tuning fork oscillators to generate stable low-frequency signals, and even sine-cosine approximators used in the input equipment for processing azimuth data.[47]

Neither these circuits nor the intricate regulated power supplies, voltage failure detectors, missing-pulse detectors, etc. will be described in detail in the following. Some additional information will be given in sections where units incorporating such devices are covered.

5.15 Pluggable units

Combinations of the basic devices described in the previous sections were mostly packaged in standard plug-in modules, called *pluggable units* as shown in figure 5.36. There were two basic variants of this basic packaging unit: One capable of holding up to nine vacuum tubes and one designed for six tubes. The mounting of the tubes is particularly noteworthy: The front plate of such a module has holes for the tubes which exhibit a larger diameter than the tubes themselves. Thus cold air supplied from the air conditioning equipment entering the computer's frames will escape through these gaps forming a cooling sheath of air around the tubes. The necessary passive components like resistors, capacitors, diodes, and coils were mounted behind the tubes on both sides of phenolic base cards – forerunners of today's *printed circuits boards*.[48] These cards were populated with the passive components and then dip-soldered during the manufacturing process. Compared with other digital computer modules from the same time frame, these modules look rather bulky and do not exhibit a density as high as similar modules of commercial computers like the IBM 701. The reason for this was maintainability: Tubes could be changed without the necessity of removing such a pluggable unit. More complex faults could be solved in a matter of minutes at most, by swapping a defective unit with a known good one. The low mean time to repair was very well worth the price of a larger footprint of the overall machine.

Figure 5.37 shows a module being repaired. Each AN/FSQ-7/8 installation featured repair workbenches like that shown in figure 5.38 which contained the necessary power supplies for operating single modules, as well as measurement equipment including a *Vacuum Tube Voltmeter (VTVM)* and an oscilloscope. External connections could be made on a patch field which contained a jack for every connection of the module as well as some signal sources like standard pulses and levels etc.

---

[46] These would be called DACs today.
[47] These were basically diode function generators, see [ULMANN 2013][pp. 77 ff.].
[48] *PCB* for short.

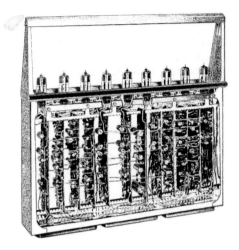

Figure 5.36: Typical nine-tube plug-in module (see [IBM BASIC CIRCUITS][p. 7])

Figure 5.37: Plug-in module undergoing service (see [Armed Services Press 1979][p. 10])

Figure 5.38: Front view of the module tester (courtesy of Mr. Ron Brunell)

5.16 The FETRON

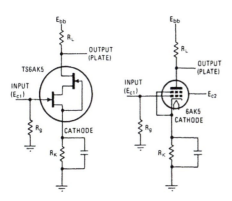

Figure 5.39: FETRON as a replacement for a 6AK5 pentode (see [Teledyne 1972] and [Burman 1972])

Figure 5.40: Production steps of a FETRON (see [Teledyne 1973][p. 4])

5.16 The FETRON

An interesting, yet short-lived development began in May 1977 at McChord AFB: A $ 100,000 testing program was set up to determine if the life-time of its AN/FSQ-7, affectionately called "CLYDE" by its staff, could be extended while reducing the overall operational costs by replacing a substantial part of its vacuum tubes, mainly pentodes, with a solid state device known as *"FETRON"*.[49]

Development of the FETRON started in early 1970 at Teledyne Semiconductor with the aim of creating a pin-compatible semiconductor-replacement for vacuum tubes which were still used abundantly in various equipment and caused increasing maintenance efforts in those years.[50]

The FETRON is based on the junction field-effect transistor – the simplest field-effect transistor of all which already exhibits a vacuum-tube-like behavior. Figure 5.39 shows the simplified schematic of a FETRON suitable as a replacement for 6AK5 pentodes: At its heart is a cascode configuration of two JFETs, the lower left one determines the input characteristics of the device while the upper right JFET increases the ability to withstand the high voltages found in typical vacuum tube circuits. Not shown in the schematic is a tantalum fuse in the plate line to protect the surrounding circuitry in case of a failure.[51] Figure 5.40 shows the various production steps of a FETRON: The JFETs are mounted on a thick film substrate forming a hybrid circuit which also holds the tantalum fuse and optional RC-networks to tailor the device's performance to that of a specific vacuum tube to be replaced. The connectors of the FETRON protrude through its bottom plate and allow a one-to-one replacement of vacuum tubes under ideal circumstances.[52]

---

[49] See [SKYWATCH].

[50] Accordingly, Teledyne offered conversion kits aimed at specific devices like oscilloscopes, radio transmitters and receivers, VTVMs etc., see [Teledyne 1973].

[51] See [Teledyne 1973][p. 5].

[52] [Holden 2011] describes a FETRON based radio designed and built in 2007 as a demonstration unit showing the unique properties of these devices in a practical example.

CLYDE's circuitry proved to be more challenging than other areas of application where FETRONs fitted in rather easily. About 9,000 diodes had to be replaced by more modern types to allow the replacement of more than 40,000 vacuum tubes. It was estimated that the break-even point regarding the expenses for this massive retrofit including the vast amount of manual labor would be reached after only 19 months of operation due to savings in power, air conditioning and lower maintenance efforts.[53]

As successful as this conversion of CLYDE proved to be – eventually, about 70% of CLYDE's tubes were replaced by FETRONs under a $ 400,000 contract[54] – it had no substantial effect on the life-time of the remaining installed AN/FSQ-7 base – the last machines were shut down for good in 1983.

5.17   Troubleshooting

Troubleshooting a machine as complex as a duplex AN/FSQ-7 central computer is by no means an easy task and would have been impossible without the aid of marginal checking and intricate diagnostic software routines. MIKE LOEWEN, a former computer maintenance technician remembers:

> "The final part of [the] training program involved pulling a random tube from the standby system, and having me deduce its location using the Duplex Maintenance Console (DMC) and available diagnostics procedures."[55]

In some cases, troubleshooting could become a nightmare, especially when the problem was intermittent, as DAVID E. CASTEEL remembers:

> "In a more serious vein, we had a problem with the B computer for several months. Although it always passed its maintenance checks, when it was active and processing the Air Defense picture, every so often very strange things would suddenly happen: Sometimes the displays would just fly to the edges of the Situation Display; sometimes features that had been selected to operate would just drop out of operation; or the converse – features that had not been selected would suddenly begin functioning. For a couple of months no one could find what was causing these strange things to happen. Eventually, [it turned out] that at least some of problems were associated with a pick or drop of one particular bit [...] – the features starting or stopping were all associated with that bit in a [control]

---

[53]See [SKYWATCH]: The annual savings were estimated being about $ 736,000 due to lower maintenance including replacement tubes, $ 54,000 of savings in air conditioning, and $ 390,000 for electrical power.

[54]Tubes had not only become rather expensive in the late 1970s but also hard to obtain. According to numerous rumors, it became necessary in some cases to buy replacement vacuum tubes for supporting the remaining AN/FSQ-7 systems, which ran until 1983, from countries of the former Soviet Union! Given the sheer number of tubes in a duplex AN/FSQ-7 computer, the required replacement parts for a single week were quite impressive, as [BELL 1983][p. 14] remembers: *"Each week they regularly replace 300 tubes and an additional 5 tubes that are showing signs of deterioration."* So replacing as many of these delicate vacuum tubes by FETRONs was a suitable alternative.

[55]See http://www.smecc.org/sage_a_n_fsq-7.htm, retrieved 12/23/2013.

5.17 Troubleshooting

word [...] We organized a plan to remove and replace every one of the pluggable units in the B computer associated with movement or changing that particular bit (dozens of them) and each was marked with where it had been mounted. Then over time we put each of those units back into its original position and left the machine to run. Eventually, we reached a point where the problems returned and thereby could identify the defective unit. Close examination revealed that there was a resistor with a hairline crack in it that sometimes was in contact and other times was not, producing spikes in the signal passing through it. That resistor was replaced and the problems stayed fixed."[56]

Despite the unique challenges of maintaining these intricate computers, the reliability achieved was (and still is – even from today's perspective) extremely remarkable: [BELL 1983][p. 14] quotes an uptime of 99.83% for a simplex machine and 99.97% for duplexed operation. Detailed performance and availability figures were collected for the last seven AN/FSQ-7 installations from March 1978 to February 1980:

"The average percentage of time that both machines of a [duplex] system were down for maintenance was 0.043 percent, or 3.77 hours per year. The average percentage of time both machines were down for all causes, including air conditioning and other situations not attributable to the computers, was 0.272 percent, or 24 hours per year."[57]

---

[56] DAVID E. CASTEEL, Captain, USAF (ret), personal communication.
[57] See [ASTRAHAN et al. 1983][p. 349].

6 Central processor

The AN/FSQ-7 central processor consists of a number of interconnected elements as shown in figure 6.1. Basically, it is a classic VON NEUMANN-architecture with a central memory subsystem holding instructions as well as data. Depending on perspective, the AN/FSQ-7 can be seen with some justification as a 16 bit computer or a 32 bit computer: The memory word-length is 32 bits plus a single parity bit for error checking purposes, while the basic operations of the computer itself work on 16 bit values in one's complement. It has been designed this way because the air defense application required a high amount of coordinate transformations. Accordingly, it was decided to implement an arithmetic unit[1] capable of working on two 16 bit values at a time. So while single data words are restricted to 16 bit in length, the AE, consisting of a *left* and a *right AE*, processed two such quantities at a time thus requiring a 32 bit memory data path. Correspondingly, a memory word is divided into a *left half-word* and a *right half-word* of 16 bits each.[2] Such a half-word x has the same structure as the machine word of Whirlwind as shown in figure 3.1, section 3.1. It is interpreted as a value $-1 + 2^{-15} \leq x \leq 1 - 2^{-15}$.

The basic timing unit is the 6 μs *memory cycle* denoting the time necessary to read or write a word from or to memory.[3] One to three such memory cycles make up a so-called *instruction cycle*, the time necessary to execute a single instruction. In this case these cycles are called *machine cycles* to distinguish them from the timing units required for pure memory accesses. While some instructions, like transferring a value from one internal register to another, can be completed in a single memory cycle, other instructions require up to three cycles to fetch or write operands etc. These are denoted as *Program Time (PT)*, *Operate Time A (OT$_A$)*, and *Operate Time B (OT$_B$)* respectively, see figure 6.2.

A machine cycle is subdivided into twelve *timing pulses*, TP 0 to TP 11,[4] which control the generation of either *instruction pulses* controlling the execution of an instruction (PT, OT$_A$, OT$_B$), or *breakout* or *break-in pulses* which control the transfer of data from memory to an output unit or from an input unit to memory.[5] Quite similar to the control of Whirlwind,[6] the actual execution of an instruction does not start with TP 0 but with TP 7. TP 0 to TP 6 belong to the preceding instruction, allowing an overlapped instruction fetch from memory. Accordingly, a single memory cycle operation

---

[1] An *AE* in the terms of AN/FSQ-7.
[2] This definition of a *half-word* is consistent with the later usage of this term by IBM.
[3] Rumor has it that IBM initially missed this 6 μs design goal since the memory did not operate reliably below 6.3 μs which accordingly became the base of the computer's cycle time.
[4] Standard pulses of 100 ns duration, with a repetition rate of 2 MHz.
[5] See section 6.3.
[6] See figure 3.15 in section 3.1.2.

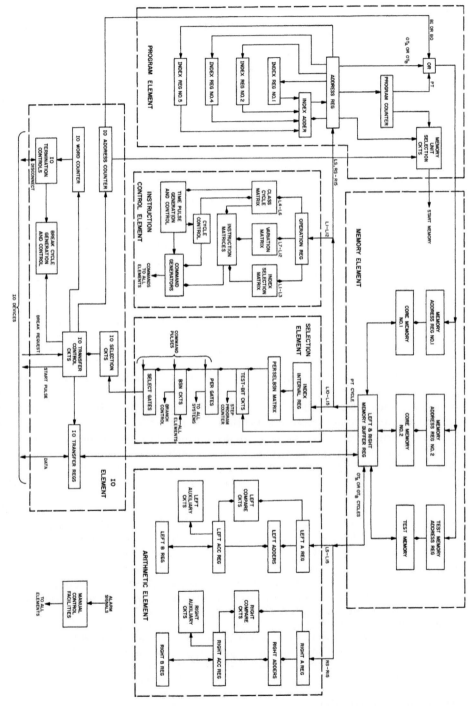

Figure 6.1: Basic circuit groups of the AN/FSQ-7 (see [IBM CCS 1][pp. 235 f.])

PT	Instruction decoding and start of execution
OTA	Fetch operand and perform execution
OTB	Write result to core memory

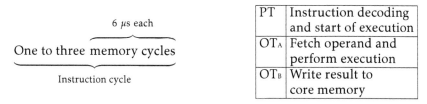

Figure 6.2: Memory/instruction cycles and cycle types

	TP 0	TP 1	TP 2	TP 3	TP 4	TP 5	TP 6	TP 7	TP 8	TP 9	TP 10	TP 11
Memory cycle	Transfer address			Read data					Write data back			
PT	Finish prev. instr.							Decode instr.		Execute instruction		
PT	Select mem. & transfer instr.								Indexing			
OTA	Select mem. & transfer operand							Execute instruction				
OTB	Result to memory buffer								Write to core memory			

Figure 6.3: Typical sequence of machine cycles during instruction execution (see [IBM CCS I][p. 90] and [IBM CCS XD][p. 29])

will start at TP 7 and will end at TP 6 of the next machine cycle. Since the following instruction has already been fetched during TP 0 to TP 6 of this cycle, its execution starts with the next pulse TP 7.

Figure 6.3 shows a typical sequence of timing pulses controlling the execution of an instruction. The first eight pulses used for the instruction fetch look like this:

TP 0: This is actually PT 0 and causes the memory address register[7] to be cleared in preparation of the following transfer of a new memory address.

TP 1: The memory buffer registers are cleared and the contents of the program counter are transferred to the memory address register initiating a read access which takes four pulse times to complete.

TP 6: The operand-address and operation registers[8] are cleared.

TP 7: The instruction read from memory is transferred to the operand-address (right half-word of the memory buffer) and operation registers (left half-word);[9] the program counter is incremented to prepare for the next instruction fetch. Since readout is a destructive operation in core memory systems, the remaining cycles TP 7, ..., TP 11 are used by the memory subsystem to write the word just read back to the core stack.

Some instructions, like multiplication, division, and shifting required more time than the 18μs of three consecutive memory cycles. These were implemented by suppressing the generation of timing pulses for a predefined duration. During this *arithmetic pause*

[7] And the *test memory* address register, see section 6.1.

[8] The *instruction register* in modern parlance.

[9] AN/FSQ-7 is a single-address machine since an instruction could contain the address of one operand.

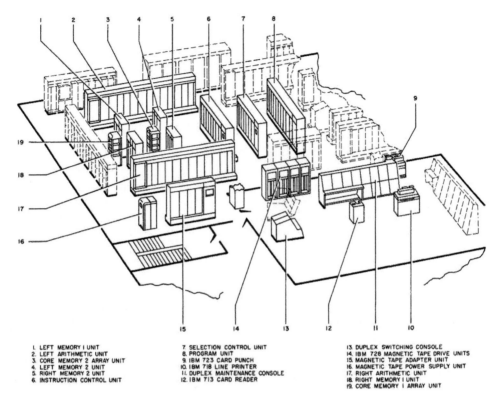

1. LEFT MEMORY I UNIT
2. LEFT ARITHMETIC UNIT
3. CORE MEMORY 2 ARRAY UNIT
4. LEFT MEMORY 2 UNIT
5. RIGHT MEMORY 2 UNIT
6. INSTRUCTION CONTROL UNIT
7. SELECTION CONTROL UNIT
8. PROGRAM UNIT
9. IBM 723 CARD PUNCH
10. IBM 718 LINE PRINTER
11. DUPLEX MAINTENANCE CONSOLE
12. IBM 713 CARD READER
13. DUPLEX SWITCHING CONSOLE
14. IBM 728 MAGNETIC TAPE DRIVE UNITS
15. MAGNETIC TAPE ADAPTER UNIT
16. MAGNETIC TAPE POWER SUPPLY UNIT
17. RIGHT ARITHMETIC UNIT
18. RIGHT MEMORY I UNIT
19. CORE MEMORY I ARRAY UNIT

Figure 6.4: Detailed, partial view of the second floor of a SAGE building, housing the central computer system (see [IBM AN/FSQ-7][p. 36])

all other central computer activities were paused. A multiplication took 32 timing pulses to complete, corresponding to 16 μs while a division instruction required 104 pulses (52 μs).

The following sections will now focus on the various subsystems of the central computer system as shown in figure 6.1: The memory element, the instruction control element, the program element, the arithmetic element, the selection element, the input/output element, and finally, the various manual control facilities necessary to control the duplexed computer system, the power supply and marginal checking system. Figure 6.4 shows the arrangement of the central subsystems of the computer itself on the computer room's floor.

6.1 Memory element

The so-called *memory element* was based on the core memory system developed for Whirlwind. Figure 6.5 shows a typical core plane as used in the early core memory systems for AN/FSQ-7. This plane contains $2^6 \times 2^6 = 4096$ individual cores. 36 such

6.1 Memory element

planes were stacked above each other, but only 33 of these were used for a 32 bit memory word and parity – the remaining three planes were used as spares and could be used to replace failed planes by rewiring the memory stack without the necessity of dismantling and rebuilding the stack as a whole.

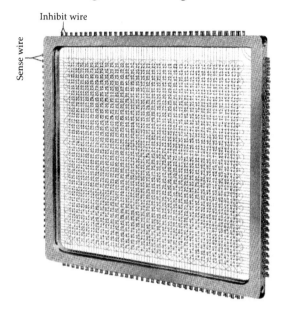

Figure 6.5: A single 64-by-64 core plane (see [IBM MEMORY ANALYSIS][p. 5])

Figures 6.6 and 6.7 show a complete memory element from its back (wiring) and front (plug-in) side. The shower stall containing the central core stack as well as the necessary 64 row and 64 column line drivers – clearly visible as groups of 16 tubes above and below the stack on all sides – can be seen in figure 6.6.

One driver out of each of these two groups of 64 driver circuits was selected by the two six bit halves of a twelve bit address by means of a diode decoder circuit similar to that shown in figure 3.12 in section 3.1.2.[10] In addition to these drivers, 33 so-called *digit plane drivers* are necessary to control the inhibit wires of the 33 planes of the core stack. In addition to this, the 33 sense wires require 33 differential amplifiers.

XD-1 and XD-2 were originally equipped with two such memory elements for a total of 8192 words of 32 bits plus parity of memory. It soon became clear that this was not sufficient for the solution of the air defense problem as HERBERT D. BENINGTON remembers:

> "[W]hen we had the XD-1 (the prototype of the AN/FSQ-7 computer) operating and had 8000 words of core[,] I started realizing then that we couldn't get the job done because there would have to be so much paging in and out from drums that we'd spend too much of our available time doing that. I was also having lunch with my boss that day, and I told him my conclusions. JAY [W. FORRESTER] dropped by at lunch and said, 'Well we've been developing a 65,000-word core memory, so we'll put it in.' That eightfold increase made the program possible."[11]

---

[10] To simplify circuitry, only five of these six bits each were actually decoded by a diode matrix driving two groups of 32 line drivers each – one group for even and one for odd addresses. Using the least significant address bit only one of these two groups was then activated.

[11] See [TROPP et al. 1983][p. 377].

Figure 6.6: 4 k Core memory system, wiring side (see [IBM MEMORY ANALYSIS][p. 4])

Accordingly, one of the 4 k memory elements was replaced by a 64 k memory element,[12] a massive and complex retrofit which affected nearly every part of the overall computer system due to the longer addresses required.[13] Figure 6.8 shows the core stack enclosure of such a 64 k memory system. With one 64 k and one 4 k memory a 17 bit address was necessary while the system was originally designed for a 16 bit address. Thus, all parts of the computer involved with address handling had to

---

[12]The 64 k memory system not only required a new memory test program, called *BIGMEM*, but also featured four spare core planes (see [MIT 1956][p. 10]).

[13]Even the program counter had to be extended since only 13 bits were initially implemented in XD-1 and XD-2, see [IBM CCS XD][pp. 219 f.].

6.1 Memory element

Figure 6.7: 4 k Core memory system, plug-in side (see [IBM MEMORY ANALYSIS][p. 3])

Figure 6.8: 64 k core memory array (courtesy of Mr. PETE KARCULIAS)

be modified. As a part of this change the AE was modified adding a special mode in which both 16 bit halves of the arithmetic element could work together on a 17 bit value instead of working in parallel on different data at a time.

Eventually, two systems, the XD-1 which was used by System Development Corporation (SDC) for software development, and the AN/FSQ-7 installation at Luke AFB, were even further extended by replacing the remaining 4 k memory element by a second 64 k element. This was necessary as Luke AFB became the primary programming site and thus required more memory than the other DCs.

In addition to this core memory element, AN/FSQ-7 featured a so-called *test memory*. It contained 16 read-only memory locations of 32 bits each which could be programmed manually using a patch panel similar to those used in contemporary punch card equipment. In addition to this, the test memory embraced two manual *toggle switch registers*, each consisting of 16 switches mounted on the maintenance console, and one 16 bit flip-flop register. The latter register was the only memory location in test memory which could be written under program control. Of these 19 memory locations, only 16 could be addressed at one time. Using four *control hubs* on the patch panel, basically every combination of these 16 read-only memory cells, the two toggle registers and the flip-flop register could be mapped to this 4 bit address space comprising the test memory.[14] The test memory was typically used during maintenance operations to store small test programs. These programs would normally test the core

---

[14]See [IBM CCS I][pp. 35 f.].

memory system and after this had been found fully operational, larger and more intricate test programs could be run from core memory, testing the overall computer system.

6.2 Instruction control element

The *instruction control element*[15] resembles what would be called a microprogrammed control unit today. It has been aptly described as being *"the internal control center of the computer since it contains the master-timing and control circuits which are necessary to sequence, co-ordinate, and control all internal operations required to execute the stored program."*[16]

As shown in figure 6.9, it is controlled by the bit pattern representing an instruction as read from core memory. The heart of the instruction control element is the so-called *DC level generator* which consists of a number of diode matrices receiving individual parts of an instruction.

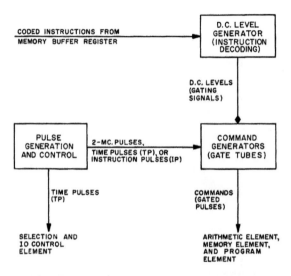

Figure 6.9: Principle of operation of the instruction control element (see [IBM CCS XD][p. 63])

Each AN/FSQ-7 instruction is specified by a *class code* denoting the class (like add, shift, etc.) the instruction belongs to, a *variation code* specifying the particular variant of the instruction, and several *index bits* controlling which – if any – of the available index registers is to be used for address calculations. Accordingly, the instruction control element contains three decoder matrices.[17] These matrices, in turn, generate a variety of DC level output signals controlling gate tube circuits.[18] These gates also receive fixed timing pulses at their respective pulse inputs. The output pulses of these circuits are then used to control the various components of the computer system. Accordingly, these gate tube circuits are called *command generators*.

Figure 6.10 shows one of these decoding matrices, the *variation matrix*, in detail: Four bits of the *operation register* are decoded by means of a number of diode AND gates

---

[15] Unit 4 in figure 4.10.

[16] See [IBM CCS I][p. 33].

[17] To save diodes and gates, not all possible instruction codes were fully decoded, which led to so-called *illegal instructions*, i.e. instructions which not explicitly implemented but resulted from those incomplete control matrices. In some cases these illegal instructions which were not trapped or suppressed otherwise proved to be useful. An example of this is any illegal instruction from the store class: It will just clear the destination memory cell.

[18] See section 5.8.

6.2 Instruction control element

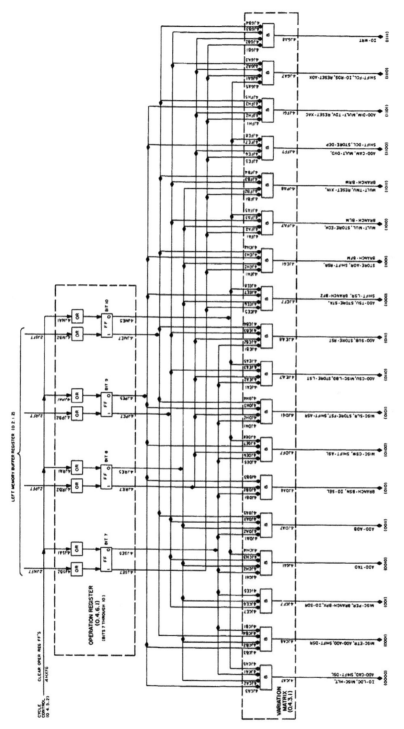

Figure 6.10: Implementation of the variation matrix of the instruction control element (see [IBM CCS XD][p. 75])

connected to the normal and inverted outputs of the register's flip-flops. All in all, the instruction control element is capable of issuing 162 different so-called *commands* in order to execute the 59 different AN/FSQ-7 instructions.[19]

Having a look at the detailed floorplan of an AN/FSQ-7 installation shown in figure 4.10 reveals that the instruction control element, unit 4, occupied a frame holding eight logic bays, each capable of holding up to 20 plug-in modules and one power distribution section. This frame contained 40 nine-tube modules of the type shown in figure 5.36, and 117 six-tube plug-in modules.

6.3 Selection and IO control element

The *selection element*[20] extends the instruction control element as it takes care of six special machine instructions which differ from the remaining instructions in that they control or sense external circuitry. These instructions are the following:

BSN: *Branch on sense* – depending on the current state of a so-called *sense unit* a conditional branch can be executed, altering the program flow.[21]

PER: Short for *operate*, this instruction controls *operate units* like condition lights, intercommunication equipment, interlock logic, magnetic tapes, line printers, card punches and readers, marginal checking equipment, etc.[22]

TOB and TTB: The *test one bit* and *test two bits* instructions compare one bit or two consecutive bits in a 32 bit word stored in core memory with a specified pattern and conditionally skip the following instruction.[23]

SEL and SDR: The *select* and *select drum* instructions select input/output devices and drums for subsequent data transfers to and from the central computer.[24]

Figure 6.11 shows the simplified block diagram of the selection element: It is controlled by the six so-called *index interval* bits of the instruction currently being executed and generates the necessary control signals for selected devices etc. The *PER SEL BSN-matrix* is a diode decoder matrix similar to those used in the instruction control element.

The *IO control element* is part of the input/output subsystem of AN/FSQ-7 and provides the actual interface to input/output devices under control of the select instructions (SEL and SDR) and the instructions to load the input/output address counter (LDC,

---

[19]See [IBM CCS I][p. 65]. The early incarnations of the XD-1 and XD-2, featuring only 48 basic instructions, required 151 such commands to be issued from the instruction control element, see [IBM CCS XD][p. 63].
[20]See [IBM CCS I][pp. 177 ff.] and [IBM CCS XD][pp. 249 ff.].
[21]See [IBM CCS I][pp. 189 ff.].
[22]See [IBM CCS I][pp. 179 ff.].
[23]See [IBM CCS I][pp. 195 ff.].
[24]See [IBM CCS I][pp. 201 ff.].

6.4 Program element

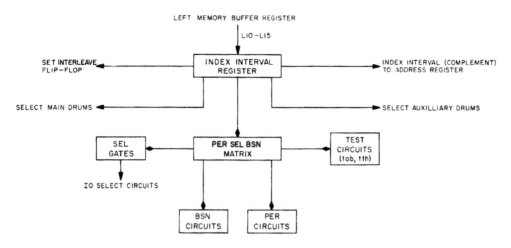

Figure 6.11: Block diagram of the selection element (see [IBM CCS I][p. 177])

load IO address counter), and to read (RDS) or write (WRT) data from and to input/output devices. In contrast to most of its contemporary computer systems and even many of today's computers, AN/FSQ-7 did not support direct input/output in which data is transferred synchronously under program control between the computer system and some input/output device.

Instead, such outbound transfers were implemented as *break-in* (from an external device to the central memory) or *breakout* transfers. To initiate such a transfer, address information must be supplied and an input/output element must be selected. Following this a *break cycle* will be initiated, causing the central computer to suspend operation for a 6 μs cycle, during which an automatic transfer of one data word takes place.[25] Accordingly, the IO control element would be called a *direct memory access*[26] *controller* today. It should be noted that AN/FSQ-7 did not support any interrupt processing at all, so it was not possible to interrupt a running program by an external event to perform some interrupt service routine. These break-cycles only halted a program for a single memory cycle after which program execution continued seamlessly.

6.4   Program element

While other digital computers of the 1950s had only one central control unit taking care of instruction decoding and execution, as well as of controlling the overall program flow, these tasks were handled by two different units in AN/FSQ-7. In addition

---

[25]*"I/O operations start block transfers of data to/from drum buffers that proceed in parallel with further CPU operations. A controller generates the sequential memory addresses for the block and decrements a counter, while the CPU has a conditional branch to test completion of the transfer. Transfers are interlocked so that the CPU is stalled if a second transfer is attempted before the previous one ends."* See [SMOTHERMAN 1989][p. 8].

[26]*DMA* for short.

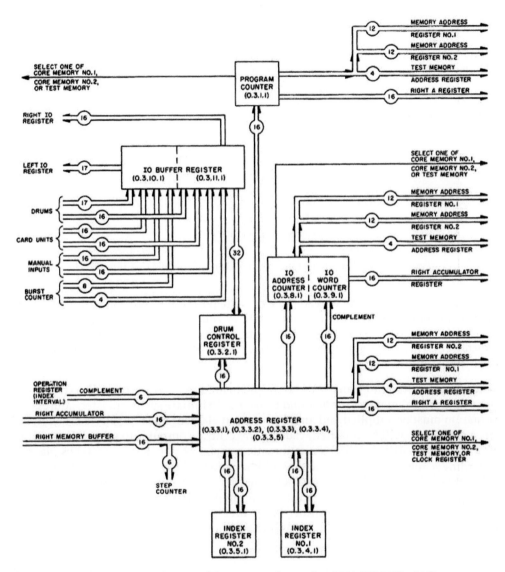

Figure 6.12: Block diagram of the program element (see [IBM CCS XD][p. 211])

to the instruction control element it featured a so-called *program element*[27] being responsible for selecting memory addresses for reading instructions and data, keeping track of the overall program flow, coordinating input/output operations and indexed addressing. Therefore the program element not only contains the *program counter* but also address counters for input/output operations, input/output buffer registers, and index and address registers as shown in figure 6.12.

---

[27] See [IBM CCS XD][pp. 205 ff.] and [IBM CCS I][pp. 159 ff.].

6.5 Arithmetic element

The *address register* receives data from a variety of sources like the address part of an instruction word, from *index registers*, or the right accumulator. Its main purpose is to supply address information to the memory system and the program counter in case of branches. It also furnishes address information for accessing the magnetic drums. The *IO address counter* and the *IO word counter* shown in figure 6.12 are used for break-in and breakout transfers resembling today's DMA transfers. Not shown is the *index adder* which is used for computing a memory address from the contents of the right accumulator and one of the available index registers.

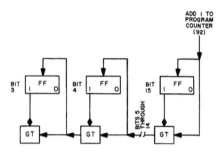

Figure 6.13: Implementation of the program counter (see [IBM CCS XD][p. 220])

The original implementation of the program counter of XD-1 and XD-2 is shown in figure 6.13. Since these machines had two 4 k core memory systems, a 13 bit program counter was sufficient and, accordingly, only bits 3 to 15 had been implemented. The individual address bits are implemented as flip-flops, each feeding the conditioning input of a gate tube circuit. Since these gate tubes are conditioned by the 1-output of the flip-flops, they effectively propagate any carry signals generated during one increment operation.

Figure 6.14 gives an overall impression of the instruction decoding process showing the interplay of all elements discussed so far.

6.5 Arithmetic element

The *arithmetic element*[28] was the heart of the AN/FSQ-7 computer – it implements all of the basic data manipulations operations like add, subtract, multiply, divide, shift, and many more. AN/FSQ-7 had to perform a lot of coordinate transformations from polar coordinates as delivered from the radar stations to Cartesian coordinates as required by the display subsystems. In contrast to most scientific digital computers of the 1950s, featuring long machine words of 36 or 40 bits, a short word length of 16 bit was sufficient for this type of processing given the limited precision of incoming radar data. To speed up the data processing, two such 16 bit words were stored in one 32 bit memory cell. Accordingly, the arithmetic element was implemented as two 16 bit arithmetic units working in parallel on both halves of a 32 bit word.

Figure 6.15 shows the block diagram of the arithmetic unit: Each half is quite similar in structure to the arithmetic element of Whirlwind and consists of an A-register (AR), a B-register (BR), and an accumulator (AC) for storing data. The central data processing element is the adder which is connected to the AC and the AR. Both halves can exchange data by shifting single bits from the left accumulator to the right or vice versa. The left and right memory registers shown at the very top of figure 6.15 are actually part of the arithmetic element to reduce the overall wire count and to keep

---

[28] See [IBM CCS XD][pp. 133 ff.] and [IBM CCS I][pp. 113 ff.].

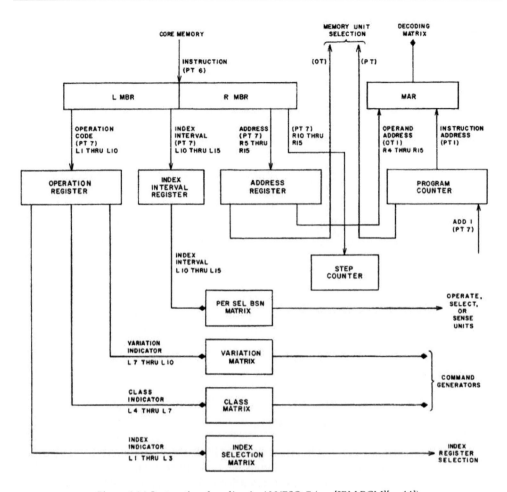

Figure 6.14: Instruction decoding in AN/FSQ-7 (see [IBM PGM][p. 44])

wire lengths down at a minimum. In addition to these registers, the arithmetic element contains left and right *IO registers, test registers*, and a *clock register*.[29]

The implementation of a single bit *full-adder*[30] is shown in figure 6.16. The left and right adders each contain 16 of these building blocks as well as a lot of other equipment to implement other instructions like shift, multiplication and division. The two flip-flops at the left hand side of figure 6.16 hold two corresponding bits of the AC and the AR of one side of the arithmetic element. Their inverted and non-inverted outputs drive four AND gates corresponding to the four possible bit combinations during an add operation. Since 1+0=0+1, two of the AND gates feed an OR gate, so that this first stage of the adder generates three mutually exclusive output signals.

---

[29]The single clock register was associated with the right arithmetic element.

[30]A full-adder adds two bits and a carry bit from the next least significant adder and yields a sum bit as well as a carry bit for the next stage. [TAYLOR et al. 1952] contains some in depth-information about the design of the AE.

6.5 Arithmetic element

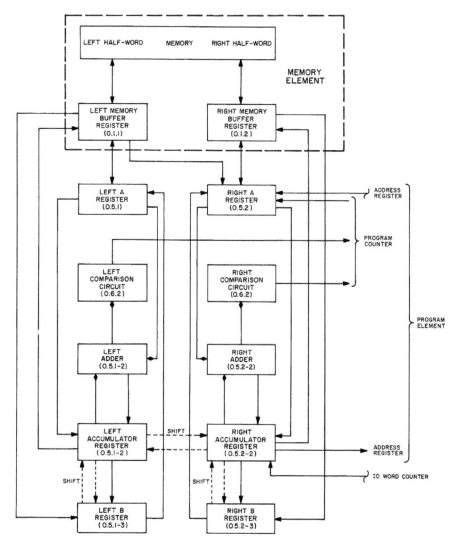

Figure 6.15: Block diagram of the arithmetic element, left and right half (see [IBM CCS I][p. 114])

Six gate tube circuits then combine these three output signals with the non-inverted and inverted carry signal from the preceding stage, yielding six output signals which are then combined to generate the sum and carry outputs. A short example may clarify the this circuit's operation: Assume that the upper flip-flop is set to 1 while the lower one is 0. Assume further that the preceding stage has generated a carry: Since this combination of AC- and AR-bits activates the second AND gate, the OR gate will yield the value 1 at its output. Accordingly, the two center gate tube circuits are in the conditioned state. Since a carry signal is present from the preceding stage, the right gate tube driven by the OR gate generates an output pulse. This causes a pulse only

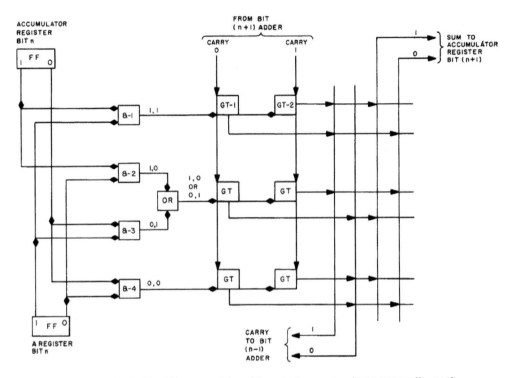

Figure 6.16: Single bit adder stage of the arithmetic element (see [IBM CCS XD][p. 144])

on the inverted sum output line denoting the sum 0. In addition to this, a pulse is also generated on the non-inverted carry output line, denoting a set carry bit for the following adder stage.[31]

To get an impression of the complexity of the left and right arithmetic element, figure 6.17 shows the block diagram of the circuitry necessary to implement the instructions ADD, SUB, TAD (*twin add*), and TSU (*twin subtract*).[32] Similar to Whirlwind, the B-register is used for shift operations – its sign bit portion is shown on the far right, connected with the accumulator and the adder stages. As can be seen, the AC, AR, and the adder form a single functional unit with the A-register holding one source operand for an operation and the accumulator holding the other operand. The result of the operation will then be stored in the accumulator.

A more careful inspection of figure 6.17 reveals an interesting feature: The result of an add operation is written back to the accumulator, shifted one bit to the right! This highly unusual feature was implemented to speed up multiplication which relies on such right shifts during each single multiplication step. Yet, an add or subtract op-

---

[31] It is interesting to note that this is a classical implementation of a so-called *ripple carry adder* where the carry information propagates from one full adder to the next, in contrast to the more intricate carry look-ahead implementation of Whirlwind as shown in figure 3.9 in section 3.1.1.

[32] Only four of the 16 adder stages are shown explicitly and the memory buffer register is clearly visible at the top of the figure.

6.5 Arithmetic element

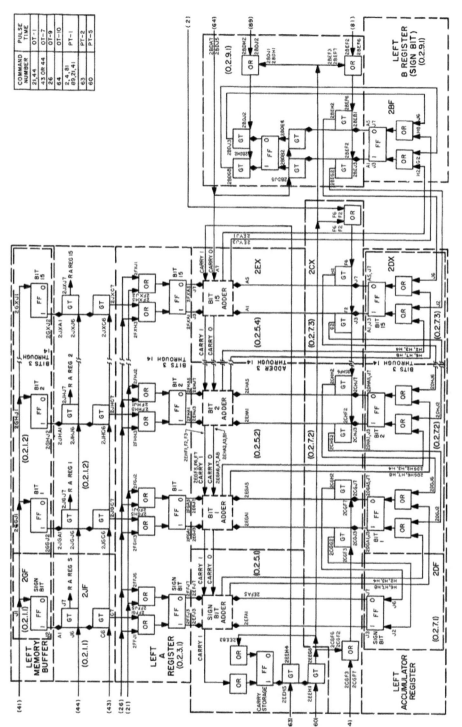

Figure 6.17: Implementation of ADD, SUB, TAD and TSU (see [IBM CCS XD][p. 163])

eration with this feature would yield an erroneous result being wrong by a factor of 2. Therefore ADD, SUB, and their variations required a corrective left shift to counteract this implicit right shift. Since this additional shift fit quite well into the timing scheme of the basic operations, no additional memory cycle was necessary so that this corrective operation did not slow down these basic operations.

Unfortunately the same does not hold true for division which needs a left shift instead of the right shift so advantageous for multiplications. While the iterative multiplication process can save one shift per iteration, division has to perform two left shifts significantly slowing down this particular instruction.[33] DAVID E. CASTEEL, Captain, USAF (ret), remembers:[34]

> "One may think that this technique would be counterproductive, but a major part of the CPU's time was devoted to making polar-to-rectangular coordinate conversions of the incoming radar data, and each such point conversion required 2 multiplications.[35] With each of up to 16 radars reporting as many as 500 data points each antenna scan (nominal 12 seconds at 5 rpm), that is a need to perform 80,000 multiplications each minute, and the designers were looking to save time any way they could. Division on the other hand, was typically used to provide the speed value of a track, and would occur much less frequently."

The complexity of the arithmetic unit is best demonstrated in figure 6.18 which shows the front and back view of the left arithmetic element: 13 bays, each holding about 20 plug-in modules, were necessary to implement the circuitry shown on the left half of figure 6.15. The leftmost bay on the front view is typical for all frames used in an AN/FSQ-7/8 and contains literally hundreds of circuit breakers. These were used to remove power from a particular plug-in module to allow replacement without having to power down the rest of the frame or even the computer. The telephone visible on the side of the frame is part of the intercom system which allowed technicians working in different areas of the large (and noisy) computer room to communicate with each other.

---

[33]That is the reason why a division required 102 or 104 timing pulses whereas multiplication only needed 34 or 35 pulses to complete. Accordingly, programmers tried to avoid division at all cost.
[34]Note to the author.
[35]To speed up operation further, the TMU instruction (short for *twin and multiply*) allowed the multiplication of two 16 bit word pairs in parallel.

6.5 Arithmetic element

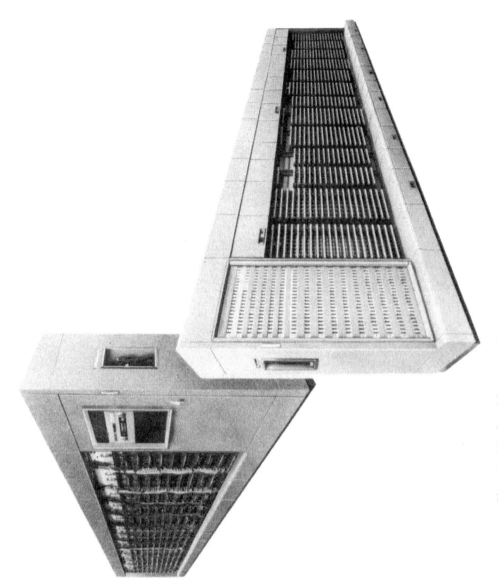

Figure 6.18: Back and front view of the left arithmetic element (see [IBM CCS I][p. 9])

7 Drum system

As pioneered in Whirlwind, the AN/FSQ-7 relied on magnetic drums to decouple the operations of the central computer system itself and the variety of asynchronously operating input/output equipment. In addition to this, magnetic drums were used as swap space for programs from core memory – another first of this remarkable machine. Figure 7.1 shows the front view of a frame holding six magnetic drums in the lower half, while the upper half is occupied by the necessary read/write-amplifiers and associated circuitry. All in all twelve magnetic drums were used in an AN/FSQ-7 installation:

Figure 7.1: Front view of the drum section (courtesy of the Computer History Museum)

LOG: The *LOG* drum served as interface between the central computer and the long-range radar input (LRI), contained the *Output Buffer (OB)*, and received data from the gap-filler radars (GFI).

MIXD: The *MIXD* drum was the interface to the manual input system (MI), *Intercommunication (IC)* (communication with the other AN/FSQ-7 computer in the DC),[1] Crosstell (XTL – communication with adjacent DCs and CCs), spare XTL, the *Digital Display (DD)*, and contained one spare auxiliary memory area where program and data could be stored and retrieved by the central computer.

RD and TD: The *Radar Display (RD)* and *Track Data (TD)* drums stored radar and track data to be displayed on the various display consoles.

AM: Eight *Auxiliary Memory (AM)* drums (AM-A, ..., AM-H) served as storage for program and data.

---

[1]"Intercommunication between duplex areas A and B of the equipment enables the alternate standby area to keep abreast of the tactical development and equipment performance in the active area. The two areas are related by means of data written by the active Central Computer System on the active Drum System IC field. This data is transferred to the standby Central Computer System under control of the standby Drum System.", see [IBM DRUM][p. 109].

In contrast to the rather simple auxiliary drum system – being quite similar to other drum systems used for intermediate program and data storage of that time – the LOG, MIXD, RD, and TD drums were much more difficult to interface as they had to decouple the central computer and the asynchronously operating input/output systems. Accordingly, one has to distinguish between so-called *Computer-Drum (CD)* transfers, where data is transmitted between the central computer and the drum, and *Other-than-computer-Drum (OD)* transfers, denoting transfers between an input/output unit and a drum, taking place simultaneously.[2]

Each of the drums is divided into *fields* 33 bits wide (32 bit data plus one parity bit) with either 2,048 memory locations per field around the circumference of the drum, or 2060 cells in the case of the RD and TD drums. The RD drum has nine such fields (72 kB) while the other eleven drums feature six fields each (48 kB).[3]

7.1 Magnetic drums

Figure 7.2 shows the layout of a typical magnetic drum used in the AN/FSQ-7. It consists of

1. the so-called *drum cradle*, a hinge allowing the drum assembly to swing out for maintenance,
2. shielding cans containing diode switches for the fields accessed through CD-transfers,
3. a 0.5 hp synchronous 3-phase drum motor running at 3600 rpm (this motor, requiring three phases at 208 V had such an excessively high starting current of 125 A that a sequencer was necessary to make sure that only one of these motors was started at a time),
4. the drum motor pulley,
5. a pulley guard,
6. drum belt,
7. shielding cans holding diode switches for OD-fields,
8. another pulley guard (transparent),
9. the drum rotor pulley (due to the transmission ratio of the motor and rotor pulleys, the drum is rotating at 2914 rpm – at this speed the average access time is 10 ms with a maximum of 20 ms, and the transfer time for one 33 bit word is 10 μs),

---

[2] The auxiliary drums are the only drums which are read *and* written by the central computer only.

[3] These fields are LRI 1/2, OB 1/2/3, and GFI on the LOG drum; MI, IC, XTL, spare XTL, DD, and spare AM on the MIXD drum; RD 1...9 on the RD drum; TD 1...6 on the TD drum; and six AM fields on each AM drum (see [IBM DRUM][p. 16]). Since an AN/FSQ-8 CC installation did not receive any radar data at all, its LRI, GFI, and RD fields were unused.

7.1 Magnetic drums

Figure 7.2: Breakdown of mechanical drum assembly (see [IBM DRUM][p. 7])

10. a static grounding brush,[4] and

11. the drum rotor itself.

The *head bars* are clearly visible located around the circumference of the drum rotor, each holding 33 read/write heads. These head bars are shifted slightly against each other in horizontal direction, resulting in a one-to-one correspondence of head bars to data fields in the case of an auxiliary drum. The remaining drums, acting as buffers between input/output equipment and the central computer, have corresponding pairs of head bars installed, so each field has two head bars – one connected to the computer and one to the associated input/output device. This arrangement allows parallel read/write transfers for a single field on these buffer drums.

Selecting a drum and a particular field on this drum is implemented by diode switches as shown in figure 7.3: Shown are two drums, A and B, with two fields each (represented by a single head per field). To select a single field, its associated *Drum Field Driver (DFD)* generates a +125 V signal at its output which is connected to the cen-

---

[4]Such grounding brushes are still necessary in modern disk drives. Without these, static electricity could build up due to friction, disturbing the sensitive read amplifiers.

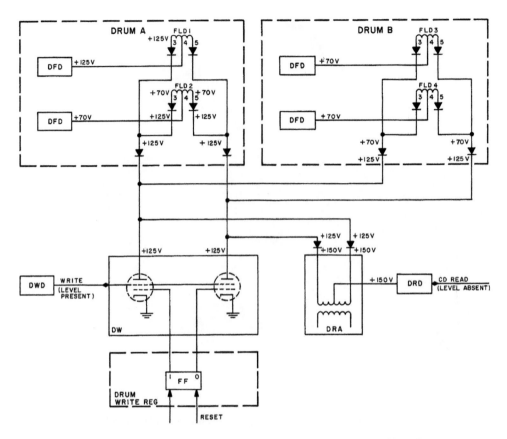

Figure 7.3: Diode switch for drum head selection (see [IBM DRUM][p. 66])

ter tap of all read/write heads belonging to that particular field. All other, unselected DFDs hold their outputs at +70 V. In the picture, the left upper field is selected – since the two diodes at the top left are now in the conducting state, as their anodes are more positive than their cathodes, the two outer connections to the selected head are connected to the plates of the two driver pentodes shown on the lower left while the other heads are deactivated.

Since the plate voltage of these two pentodes is supplied by the selected DFD via the selected head and associated diodes, all, what is necessary to write a bit to the drum, is to load the associated flip-flop in the *drum write register* with the desired value and activate the pentode pair by a positive write level at their suppressor grids. This will tie one end of the read/write head to ground potential while the other will stay at +125 V. Writing a 0 energizes one half of the read/write head while writing a 1 energizes the other half under control of the drum write register.

Reading from the drum works similarly: Each field has an associated *Drum Read Driver (DRD)* which is connected to the center tap of a decoupling transformer like that shown on the lower right in figure 7.3. An unselected DRD yields +150 V at its output,

7.1 Magnetic drums

effectively blocking the two diodes connecting the transformer with the read/write head diode network. When a field is selected for reading, its corresponding DRD output will be at +100 V, so that the two diodes at the transformer will conduct. Accordingly, the secondary of the transformer will show the read-signal from the selected read/write head, feeding a read-amplifier.

Obviously the gap between the read/write heads and the surface of the drum rotor is critical regarding magnetic flux density and thus read/write levels. Figure 7.4 shows the structure of such a read/write head: Its outer aluminum case is tightly screwed to the head bar while the read/write coil with its core and the core gap slides on a carriage within this enclosure. Using the adjustment screw at the top of the head assembly the gap between the lower tip of the actual read/write head and the drum surface can be adjusted.

Since typical run-down and restart times for these drums were of the order of 15 to 20 minutes, some technicians not only adjusted but replaced heads on a running drum as MIKE LOEWEN remembers:

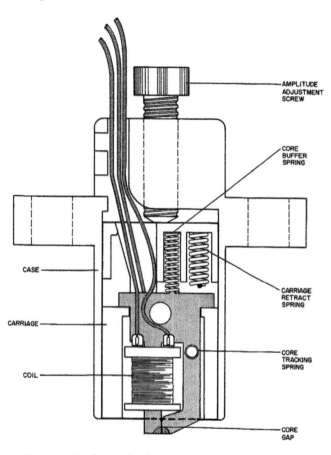

Figure 7.4: Read/write head assembly (see [IBM DRUM][p. 10])

"Some of the maintenance procedures were particularly memorable, such as changing heads on the drums. Because it took in excess of 20 minutes for a drum to spin down, you couldn't often afford the down time and you changed the head while the drum was turning at speed. It was a nerve wrecking process which involved using an oscilloscope to monitor the signal amplitude on the head. You cranked the head down towards the drum surface until the signal reached the proper level. If the head went too far and actually touched the surface of the drum, it would score the surface and make that track unusable."[5]

---

[5] See http://www.smecc.org/sage_a_n_fsq-7.htm, retrieved 12/23/2013.

Figure 7.5: Memory drum undergoing maintenance

Figure 7.5 shows a drum undergoing maintenance: A technician adjusts heads for maximum read levels guided by an oscilloscope displaying the current read signal from the particular head under adjustment.

7.2 Timing

The generation of timing signals controlling the actual read/write operations is a central aspect in every rotating storage medium. While modern disk drives normally employ timing and servo information[6] which is embedded directly in the data tracks, older systems used a dedicated track for deriving timing information. Basically all drums used in AN/FSQ-7 featured two dedicated tracks holding timing information: One track generated 2048 (or 2060 in the case of the RD and TD drums) pulses per revolution while the other track generated one pulse per revolution yielding the necessary index information denoting the first sector of a track.

---

[6]Servo information is necessary in disk drives to control the movement of the head positioning mechanism – something which is not necessary in fixed-head drums as those employed in AN/FSQ-7.

7.2 Timing

The early incarnations of the drums used a glass disk mounted to the drum shaft containing a sinusoidal pattern representing the timing information.[7] Timing information was read from this disk by an optical system containing a light source, seven lenses, and a photomultiplier tube with associated amplifiers and logic circuitry to derive proper timing pulses. Since this type of pulse generation would have required a substantial amount of additional hardware per drum,[8] the timing information derived by this optical system was written to two tracks on the drum in a preliminary step. Thus a single instance of the signal conditioning hardware, which could be switched from one drum to another, was sufficient to generate timing information and write this onto the two dedicated timing tracks.

This capability of rewriting timing and index information derived by means of this optical system with its associated electronics package was necessary, since erasing a disk, which was sometimes necessary during maintenance, erased all tracks, not just data tracks.[9]

Later,[10] another timing generation method based on an etched timing disk was used, thus rendering the complex optical system with its associated signal generation and shaping electronics obsolete. Figure 7.6 shows such a timing disk with an enlarged section showing the timing and index slots. This disk was mounted on the drum rotor shaft and could be read with a normal read head and associated electronics just like the timing channels written by the OTG in earlier devices. Due to its structure, the etched timing disk was not affected by an erase operation, simplifying operations considerably.

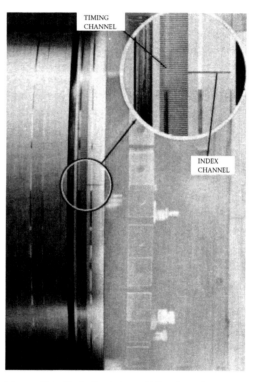

Figure 7.6: Etched timing disk of a drum (see [IBM DRUM][p. 12])

Although data on a drum could simply be overwritten by writing a new bit stream to a particular track, the drums featured a so-called *erase bar* which allowed all tracks of a drum to be erased at once. This bar was in effect just a single write head as wide as the drum rotor itself. It was powered by a motor controlled Variac during an erase operation. The output of this

---

[7]This method was used in the first 16 AN/FSQ-7/8 installations, see [IBM DRUM][p. 35].

[8]Called *Optical Timing Generator (OTG)*.

[9]In cases where all drums of a system were erased during maintenance operations, rewriting the timing and index channels on the drums could be performed automatically, controlled by a sequencing system which selected one drum after the other, making the necessary connections between the optical pickups, the timing pulse generator and the write circuits (see [IBM DRUM][p. 180]).

[10]Beginning with the 18th installation.

Variac – driven by a 125 V, 60 Hz signal – was varied slowly from its maximum amplitude to zero. This ensured that the drum did not contain any magnetization patterns at all after an erase operation.[11]

7.3 Status concept and time stamps

Since all of the drums (with the exception of the auxiliary drums) are used to decouple the asynchronous operations of the central computer and its associated input/output equipment, a synchronization mechanism is necessary.[12] Therefore, all input and output fields have two associated status channels of one bit each, the so-called *OD* and *CD status channels*. Words of a field containing valid data ready to be read by the computer system are flagged by a 1 bit in their associated CD status channel, while the OD status channel marks fields which have already been read by the computer and can be used by the input equipment to store new incoming data. The CD status channel of an output field tells the computer which words of a field are ready to accept new data while the OD status channel controls the operation of the output system.[13]

In addition to this embedded and hardware supported status scheme, timing information derived from the clock register being part of the right arithmetic element was inserted automatically into bits L10 to L14 of each GFI data word as it was transferred from the gap filler radars to the GFI field of the LOG drum. This was necessary since all calculations regarding targets and tracks rely primarily on precise time information regarding the input data.

7.4 Data flows

Figure 7.7 shows the overall data flows between the central computer system, the various magnetic drums, and the associated input/output equipment. The central role of the drum system is obvious. Basically, five types of data transfers can be distinguished:[14]

1. Data source and destination is the central computer, i. e. the drum is used to store programs or data under computer control which is the task of the eight auxiliary drums. These drums do not contain any status channels since they are operated under program control only.

---

[11] See [IBM DRUM][pp. 175 f.].

[12] See [IBM DRUM][pp. 77 ff., pp. 115 ff., and pp. 133 f.].

[13] It is interesting to note that due to the high difference in data rates between the central computer and the output system, consecutive data words were written in an interleaved way to alternating addresses of the output fields thus slowing down the data rate with respect to the output system by a factor of two. In addition to that, the XTL fields of the MIXD drum contained an additional *marker channel* which could be written by the computer to indicate the start of an outgoing multi-word message.

[14] See [IBM DRUM][p. 21].

7.4 Data flows

2. Data is written to a drum by the input system and read by the computer. Drums used in this mode are the LOG (LRI and GFI fields) and the MIXD drum (MI and XTL fields).

3. Data is written to the drum by the computer and read by the output system. This involves the LOG drum (OB fields).

4. Data is transferred from the computer to the display system via TD, RD and MIXD drums.

5. The active computer transmits data to the standby computer through the MIXD drum (IC field).

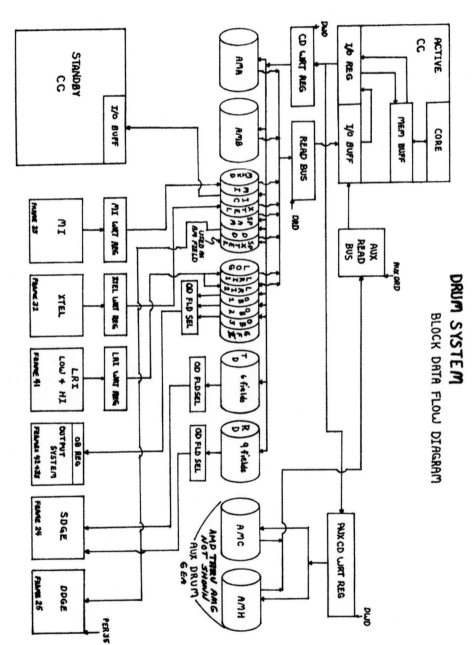

Figure 7.7: Drum system data flow (courtesy of Mike Loewen)

8 Input/output system

The input/output system of an AN/FSQ-7 installation, decoupled by the LOG and MIXD drums as shown in figure 7.7, connected the central computer to the variety of input and output channels such as radar stations, manned interceptor bases, missile bases, other DCs and CCs etc. In essence, three basic types of data transmission regarding such input and output operations can be identified:[1]

1. High data-rate sources and sinks which were directly connected to the computer by means of 1,300 bit/s channels over voice-grade telephone lines and radio channels – typical examples are radar inputs and communication links to adjacent centers.[2]

2. Lower speed channels used to connect Teletype (TTY) terminals.

3. Finally, voice communication was used for data which was either not too time-critical or where the development and use of an input/output channel would have been unnecessarily complicated or expensive. Typically such data were entered via keyboard and punched cards.

In contrast to most current computer systems, the input and output systems of a DC or CC were implemented separately. The input system was responsible for accepting data from various data sources like radar stations and writing this information to the associated drum fields. The output system, on the other hand, processed data read from its associated drums, performed some necessary preprocessing steps before so-called *messages* were assembled, serialized and sent over telephone lines to their respective destinations. The following sections describe the input and output system in more detail.

8.1 Input system

The purpose of the input system is basically to accept data delivered via telephone lines to a DC or CC. Data sources were XTL from adjacent DCs and CCs, LRI radar stations, and GFI stations. While XTL and LRI data were received as messages consisting of half-words suitable for direct processing by the central computer, the GFI radar stations transmitted SDV data requiring a substantial amount of preprocessing in the input element before it could be written to the magnetic drum system for further processing.

---

[1] See [EVERETT et al. 1983][p. 334] and [BENINGTON 1983][p. 354].
[2] The modulator and demodulator equipment used for these telephone channels is similar to that developed for Whirlwind, see section 3.6.

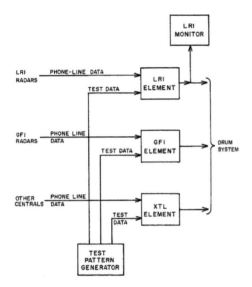

Figure 8.1: Simplified block diagram of the input system (see [IBM INPUT][p. 2])

Figure 8.2: LRI element (unit 41) during installation (see [IBM INPUT][p. 20])

Figure 8.1 shows the basic structure of the input system: As can be seen, there is no single input system – in fact, three subsystems are operating in parallel, the *LRI element* accepting incoming data from long range radar installations and providing additional means for displaying radar data prior to processing by the central computer, the *GFI element* processing incoming data from gap filler radar systems, and finally the *XTL element* which is similar to the LRI element, receiving and processing incoming crosstell data from adjacent installations.[3] These three subsystems are described in more detail in the following.

All three input subsystems could be driven by a *test pattern generator*, TPG for short, for maintenance purposes. The TPG was capable of generating signals of the same structure as those normally received from the input channels.

8.1.1 LRI element

The LRI element,[4] shown during installation in figure 8.2, received data in form of binary *messages* from adjacent LRI radar stations. Each such installation was connected to a DC using two dedicated telephone lines, one transmitting data from the radar system, while the other line was connected to the IFF equipment installed at the radar site. Since an LRI element featured 32 telephone lines plus two spare lines,[5] up to 16 LRI sites could be connected to a single DC. An LRI message, describing one potential

---

[3] The input system of an AN/FSQ-8 installation featured only an XTL element as a CC was not connected to any radar stations at all.

[4] See [IBM INPUT][pp. 19 ff.].

[5] These spare lines could be semiautomatically substituted for a regular channel in case of a failure, see [IBM INPUT][pp. 27 f.].

8.1 Input system

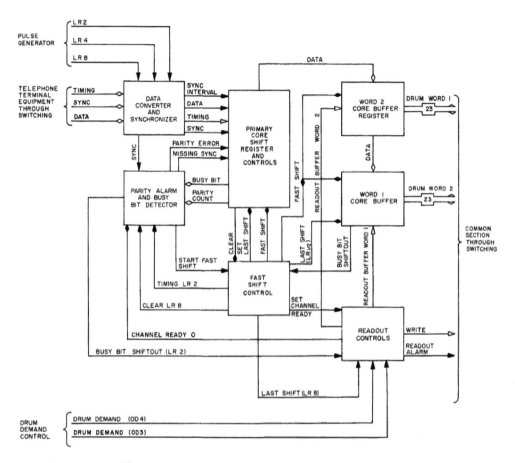

Figure 8.3: Simplified block diagram of the LRI channel equipment (see [IBM INPUT][p. 32])

target, consisted of 52 bits received at a speed of 1,300 bits/sec. Of these, 46 contained actual target information while the remaining bits were used for synchronization purposes and the like.

Figure 8.3 shows a simplified schematic diagram of the input channel equipment of the LRI element: Incoming data is fed by the *data converter and synchronizer* into a two-stage core buffer memory consisting of a *primary core shift register* and two *word core buffer registers*. This two-stage design is necessary to allow buffering of an incoming message while the previous message has not yet been written to the magnetic drum system.[6]

An LRI message is split into two words, containing azimuth and range information of a target, and is stored in the primary core shift register at a speed of 1,300 bits/sec. It is then transferred to the two word core buffers at a speed of 50,000 bits/sec. These two interconnected word core buffers feature a parallel readout for delivering data to the

---

[6]See [IBM INPUT][pp. 34 ff.].

magnetic drum system where it is written to a free address of a LRI field, a so-called *free slot* as determined by the associated status information. Prior to this, incoming data is checked regarding parity, missing or erroneous synchronization bits, etc. In case of an error the respective message is discarded.

It is interesting to note that, since the associated LRI radar installations did not transmit any data suitable for identification of a particular station, only the number of the telephone channel could be used for identification. As this was not accessible to the central computer, the LRI element added a site specific identification code based on the receiving line to each accepted message prior to storing it on the LOG drum.

Figure 8.4: LRI monitor equipment (see [IBM INPUT][p. 78])

The LRI element provided means to display incoming data on a so-called *LRI monitor*. Since incoming data was already in binary form, it had to be converted into suitable analog signals to control a PPI display like the LRI monitor. This was done by the *LRI monitor control unit*, unit 93,[7] shown in figure 8.4 together with the various display units used in this context. The LRI monitor control unit was connected to the same data lines feeding data from the LRI element to the magnetic drum system.

Various incoming messages from adjacent long range radar stations could be selected for display on the LRI monitor consoles. These displays showed the air situation as seen by the radar systems after preprocessing by the LRI element but before the central computer had processed this data. The LRI monitor consoles could be used to check data quality of incoming radar information during maintenance operations as well as to get an overview of the current air space situation even without support of the central computer. Typically one of these consoles featured a camera for documentation purposes.

---

[7] This operation required quite some analog circuitry, especially for deriving the necessary sine and cosine functions etc. to control the PPI displays, see [IBM INPUT][pp. 101 ff.].

8.1 Input system

The display units shown in figure 8.4 were based on a 16 inch display tube. Unit 623 was located in the air surveillance area and used by the air surveillance officer, the similar units 622 were installed in the maintenance area and accordingly used for maintenance purposes. Units 620 and 621, the latter featuring the aforementioned camera, were used by the mapper supervisor to get a quick overview of the air situation.

8.1.2 GFI element

While the long range radar systems already provided data in binary form suitable for processing by an AN/FSQ-7 system due to their AN/FST-2 pre-processors, the gap filler radar stations were simpler and transmitted information in form of SDV signals. Accordingly, the structure of a GFI element as shown in figure 8.5 is vastly different from that of a LRI element since a substantial amount of preprocessing was required before feeding data to the central computer.

AN/FSQ-7's GFI element allowed the connection of up to 16 incoming telephone channels plus two spare channels, each connecting one gap filler radar station to the DC. One such channel consisted of three separate telephone lines transmitting data at a speed of 1,600 bits/sec. The three lines of a channel transmitted three different types of raw data: *Azimuth* and *north azimuth* data, *range* data, and *target* information. A single revolution of the gap filler radar antenna was divided into 256 sectors – for each such sector, a pulse was transmitted over the azimuth/north azimuth channel. Every time the rotating antenna went through its north position, a double pulse was transmitted denoting the start of a new cycle. Range was divided into 64 so-called *range boxes*. Range information was transmitted as a continuous 1,600 Hz sine wave with one cycle representing one range box. The length of this signal corresponded to the actual range that had been scanned for the current azimuth sector. A target in the current sector and range box was denoted by a single cycle of a 1,600 Hz sine wave transmitted over the target line.

Figure 8.5: GFI element (unit 34), see [IBM INPUT][p. 118])

One of the preprocessing tasks performed by the GFI element and its associated equipment concerned removing *clutter*.[8] Clutter signals are caused by reflections of the radar signal from fixed landmarks etc., and had to be removed to avoid overloading the central computer. As in the early applications of Whirlwind for the air defense problem, this was done by so-called *mapper consoles*[9] as shown in figure 8.6.

These mapper consoles were actually PPI displays showing incoming raw data from the gap filler radar stations. Unwanted areas causing clutter were masked by applying an opaque paint on the screen of the display tube. On top of this tube a photomultiplier assembly was mounted which detected light from the display and generated an impulse denoting the occurrence of a relevant radar return. Up to 18 such mapper consoles, units 600 through 617, could be installed in a typical DC.

Figure 8.6: GFI mapper console [IBM INPUT][p. 118])

The basic organization of such a mapper console is shown in figure 8.7. Being a PPI display, the electron beam of the display tube is deflected by a rotating yoke which is synchronized with the rotation of the associated radar antenna. This yoke is driven by a 2-phase motor with an associated gear box and a brake system. The speed of this motor is controlled by the azimuth signals while the absolute position of the yoke is synchronized with the north azimuth pulses. The motor is driven in such a way that one revolution of its axis is performed for each azimuth pulse, so that a gear reduction of 256 is necessary as one revolution of the radar antenna is divided into 256 sectors.[10] The necessary dynamic range of this system is remarkable as the radar antenna could be driven at speeds between 2 and 10 rpm.

The gear box contains two differentials, allowing for normal and double angular velocity of the deflection yoke. When the yoke's position is lagging behind that of the radar antenna, a double-speed differential[11] is engaged for a sufficiently long interval to catch up the difference. When the yoke, driven at twice its normal speed, reaches its north position, a brake is activated, stopping the yoke assembly at this position. This

---

[8] See section 2.3.
[9] See [IBM INPUT][pp. 147 ff.].
[10] See [IBM INPUT][pp. 148 ff.].
[11] See [IBM INPUT][p. 152] for a detailed description of the differential system.

8.1 Input system

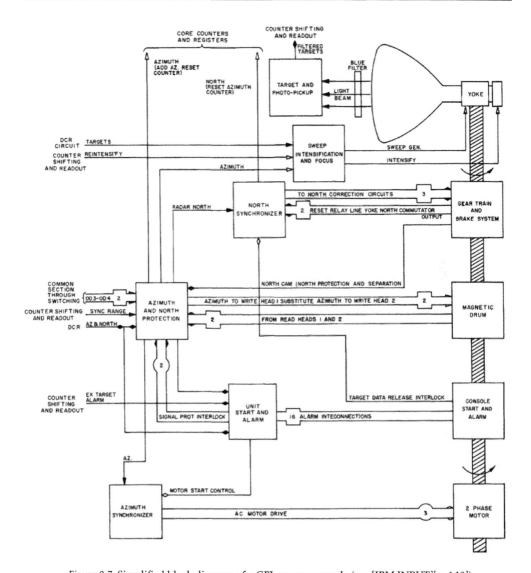

Figure 8.7: Simplified block diagram of a GFI mapper console (see [IBM INPUT][p. 148])

brake is released by the next incoming north azimuth pulse and the yoke is then driven through a single-speed differential at nominal speed. A leading yoke is also stopped at its north position and released with the next incoming north azimuth signal.[12] All of these operations are controlled by the *north synchronizer*.

A basic problem with incoming data from gap filler radar stations were spurious or missing azimuth pulses which could compromise the synchronization of the mapper consoles. Therefore some additional circuitry was necessary to suppress spuri-

---

[12]See [IBM INPUT][pp. 149 ff.].

ous pulses and to generate ersatz pulses in case of missing input signals. At the heart of this protection system is a small magnetic drum with two tracks, driven in synchronism with the deflection yoke of the mapper display tube. Incoming pulses were rejected if they were outside a small timing window, thus effectively suppressing spurious azimuth pulses. Pulses passing through this timing window were written to the two tracks of the drum, one track holding azimuth pulses while the other track stored the north azimuth pulses. Incoming pulses erased any existing information at this angular position of the drum. The synchronization process itself was then controlled by azimuth and north azimuth pulses read from the magnetic drum instead of the the incoming pulses. Since missing pulses did not erase the information recorded on the drum during the previous revolution, these replaced the missing pulse ensuring smooth operation of the synchronization circuits.

8.1.3 XTL element

Figure 8.8: XTL element during installation [IBM INPUT][p. 162])

DCs and CCs were coupled together by telephone lines carrying XTL data. This network was used to exchange information about targets and their respective tracks in other sectors.[13] A first experimental XTL connection was set up in 1955 between Whirlwind – then located in Cambridge, Massachusetts – and the XD-1 system in Lexington, proving the practicability of such an arrangement.[14] XTL data was received by the last element of the input system, the XTL element shown in figure 8.8.

The XTL element is capable of receiving incoming data from up to 22 channels plus two spare channels.[15] A single XTL message consists of five words of 17 bits each, plus six blank timing pulses and a synchronization bit for a total of 92 bits.[16]

Such XTL messages were transmitted over so-called *party lines*.[17] Accordingly, the messages had to contain address information uniquely identifying the destination center. The XTL element received the messages destined for its respective

---

[13]See [FELSBERG 1969][p. 3].
[14]See [WILDES et al. 1986][p. 300].
[15]Some DCs only supported eleven channels plus a single spare channel, see [IBM INPUT][p. 11].
[16]See [IBM INPUT][p. 3].
[17]A (multi)party line was a classic way in early telephone systems to share a single connection between more than two communications partners.

DC or CC, performed some basic error checking and stored the individual message words on free slots of the associated magnetic drum fields.

8.1.4 TPG

To test complex systems as the LRI, GFI, and XTL elements, and to perform maintenance operations, specialized test equipment was necessary. Its heart was the so-called *test pattern generator*, TPG for short, shown in figure 8.9.

The purpose of the TPG was to simulate telephone line input signals for the three input elements described above. As can been seen in the figure, it consists of three independent pattern generators, one for each input element. The leftmost element generated bit patterns suitable for driving the LRI element, the more complex pattern generator in the middle simulated the analog SDV output of a gap filler radar station, while the rightmost element generated typical XTL messages. Due to the digital nature of the LRI and XTL input channels, their associated TPGs were rather simple to implement, while the GFI TPG was much more complex as it had to generate the three signals comprising the SDV signal.[18]

Figure 8.9: TPG [IBM INPUT][p. 224])

8.2 Manual data input element

The various radar stations and adjacent XTL sources were not the only source for input data: Manual input from operators at the display consoles deciding which actions should be taken also had to be fed into the central computer. Possible input devices were *computer entry punches*,[19] *area discriminators*,[20] *light guns*, the manually operated switches at the display consoles, and track information generated by the SDGE.[21] While the LRI, GFI, and XTL elements were all coupled to the main computer through

---

[18]Since an AN/FSQ-8 installation only featured an XTL input element, it also had only a XTL TPG.

[19]This device, an IBM model 020 printing card punch and reader, was used to enter rather static data such as weather information or flight plans, see [IBM DSP 2][p. 126].

[20]See section 9.3.

[21]See section 9.1 and [IBM DSP 1][p. 151].

associated fields on the magnetic drum system, the *manual data input* element, *MDI* for short, had two ways of sending data to the computer: Data received at a rather high data rate from devices like the computer entry punch or the area discriminators was transmitted asynchronously to the computer through the MI-field of the MIXD drum, while single-bit data arriving at much lower speeds, such as generated by operators pressing keys, was stored in a dedicated magnetic core memory consisting of 128 rows of 33 cores each[22] in the manual data input unit for direct readout by the central computer. This part of the MDI element is called the *direct entry section*.

Figure 8.10: Manual data input unit (see [IBM DSP 1] [p. 5])

An operator activating a key on his display console caused one or more corresponding cores in the direct entry section to be set.[23] A so-called *keyboard control panel message* could consist of several depressed keys. The end of a message to be processed by the computer was denoted by pressing an *activate* push-button which also set a particular core in the MDI core storage. Pressing the trigger button of a light gun aimed at a particular target also set a corresponding core in this dedicated storage system. In addition to this, target data and an eight bit identification code for that particular light gun were written to the MIXD drum.[24]

The manual data input element actually consisted of two units, the *manual data input unit* shown in figure 8.10, and the *manual data interconnection unit* which was basically a large patch field holding the connections between the manual data input unit and the various manual inputs from the display consoles etc.

8.3 Output system

In addition to that intricate input system, each DC and CC featured a dedicated output system operating asynchronously and coupled to the central computer through the OB-fields of the LOG drum. The purpose of this system was to generate six groups of serial bitstreams transmitting data to various weapon systems etc. Figure 8.11 shows a

---

[22]Only 32 of the 33 cores were used for actual data storage – the remaining core was used for testing purposes and permanently set to one, see [IBM DSP 2][p. 142].
[23]Core can also be reset by releasing their associated key.
[24]See [IBM DSP 2][pp. 127 f.].

8.3 Output system

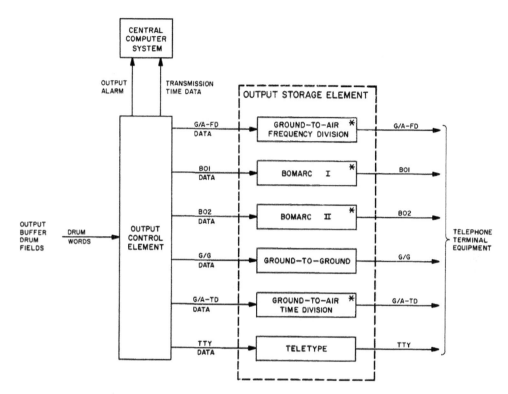

Figure 8.11: Block diagram of the output system – units denoted with an asterisk are not used in the AN/FSQ-8 (see [IBM OUT][p. 2])

simplified block diagram of this system which consists of two distinct units, the *output control element* (unit 42), receiving data from the LOG drum and the *output storage element* (unit 33), feeding the various transmission lines.[25] Data to be transmitted was written by the computer to the OD drum fields as 32 bit words (plus one parity bit). The right half-word of such a 32 bit word contained actual message information to be sent while the left half-word contained control and address information uniquely identifying the destination sites for a message.

The output system was capable of generating the following five different output formats: So-called *Ground-to-Air Frequency-Division (G/A-FD)* and *Ground-to-Air Time-Division (G/A-TD)* messages contained data supporting manned fighter aircraft to intercept targets. A G/A-FD message consisted of up to 13 right half-words, each containing a four bit interceptor address, a four bit message label and eight bits of actual data. Data was transmitted to a *data-link transmitter site* via telephone lines. From there it was transmitted by radio to the interceptors.[26] In contrast to that, a G/A-TD

---

[25] In today's parlance, the output system could be described as a serial line transmitter. In fact, the data transmission worked along the lines established during the early developments at Whirlwind, based on three parallel channels for timing, synchronization, and serial data (see [IBM OUT][p. 95]).

[26] The up to 13 message half-words were actually transmitted in parallel by 13 modulator units, each tuned to a different frequency (see [IBM OUT][p. 4]).

message consisted of two groups of two right half-words each. These two groups were transmitted over two distinct telephone lines to the manned interceptor fighters using radio data-link transmitters. One of these lines carried site, aircraft address[27] information, and message identifiers, while the other telephone line was dedicated to the transmission of the actual intercept data.

BOMARC 1 (BO1) and *BOMARC 2 (BO2)* messages were used to control unmanned interceptors, namely BOMARC missiles. One such message could contain as many as 13 words consisting of a three bit missile address, a four bit message identifier, and eight bits of data.

XTL data to adjacent DCs, so-called *forward telling data* for CCs and *height finder requests*, are denoted as *Ground-to-Ground (G/G)* messages.

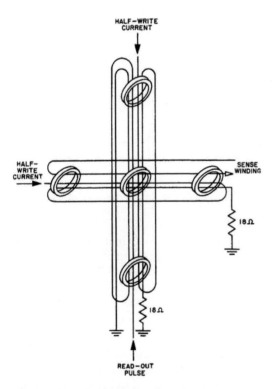

Figure 8.12: Principle of operation of the output storage core matrices (see [IBM OUT][p. 14])

The last message type generated by the output system were TTY messages. Headquarters higher in the military hierarchy, sectors lacking automatic input equipment, and antiaircraft batteries received (early warning) information through these messages.[28] Each right half-word of such a message contained three five bit TTY character codes.[29]

The principle of operation of the output storage element is noteworthy: At its heart is a modified magnetic core memory consisting of a single two-dimensional core matrix. Outgoing messages were stored in the rows of this matrix by selecting the row to be written and selecting only those columns where cores had to be set to their 1-state. Accordingly, the rows are denoted as *address axis* while the columns form the *message axis*. Due to the two-dimensional structure of this memory element no inhibit wires with their associated driver elements are necessary. The actual selection of cores to be switched to 1 is done by the traditional half-current selection scheme. Figure 8.12 shows five cores from this matrix with their associated driver and sense lines.

---

[27] Based on this address information, a particular transmitter site was selected (see [IBM OUT][p. 5]).

[28] Generally speaking, TTY messages were *"employed by the Central to carry information to remote sites not equipped with automatic input devices [...]"* (see [IBM OUT][p. 5]).

[29] An AN/FSQ-8 installation featured only the G/G and TTY output channels in its output system.

8.3 Output system

The *"HALF-WRITE CURRENT"* captions are a bit misleading given the fact that both windings cross all cores in their respective row and column twice. Due to this, the current flowing through these lines is, in fact, only a quarter of the full write current so that only the core at the intersection of one address and one message line will be switched to its 1 state.

A number of n messages $m_1, \ldots, m_n$ of i bits each were thus stored one message per row, the first row holding $m_{1,1}, \ldots, m_{1,i}$ etc. In contrast to the core memory structure used for AN/FSQ-7's main memory, the core matrices of the output storage element used a different readout method: Each column of cores had its own dedicated readout pulse running twice through all cores. To read the messages stored in a matrix, its readout lines were pulsed sequentially, starting from the left. This caused all cores of a column to be switched to their 0-state. If their previous state was 1, this change in magnetization caused a small pulse at their associated sense windings running horizontally through the matrix. Reading out all messages in this sequential manner, the sense lines carried the bits $m_{1,1}, m_{2,1}, \ldots, m_{n,1}, m_{1,2}, m_{2,2}, \ldots, m_{n,2}, \ldots, \ldots, m_{n-1,i}, m_{n,i}$. These bits were then transferred into an output shift-register driving the associated telecommunications equipment.

At first sight it looks strange to interleave the individual bits of message words to be transferred in that way, sending all first bits, then all second bits etc. The rationale behind this scheme was that the input elements at the receiving sites, including the XTL elements, performed the reverse operation by storing incoming bit sequences row-wise in a similar core matrix. These bits were then read out vertically again, restoring the proper sequence of message bits.

Figure 8.13: Core matrix of the output storage element

Figure 8.13 shows an actual output memory of an AN/FSQ-7 output element. Clearly visible are the row and column drive lines with their associated termination resistors on the top and left of the picture.

One such core matrix can store more than one message at once. Transmitting all message stored at a given instant is called a *burst*. Table 8.1 shows the maximum number of messages in an output burst for each type of output channel, their respective size, and the duration of such a burst.[30]

---

[30] The number of bits per output word listed in table 8.1 contains start and synchronization bits, although the right half-words being transmitted are always 16 bits in length. It should be also noted that the bit sequences of G/A-FD messages are reversed, compared with the other message types, since the airborne data receivers expected incoming data with different endianness.

OUTPUT MESSAGE	NO. OF MESSAGES	WORDS PER MESSAGE	BITS PER OUTPUT WORD	BURST PERIOD
G/A-FD	2	13	18	0.25 sec.
BO1	2	13	18	0.25 sec.
BO2	2	13	18	0.25 sec.
G/A-TD	3	4	17	0.09 sec.
G/G	5	5	17	0.07 sec.
TTY	25	1	18	0.495 sec.

Table 8.1: Maximum number of messages and size of messages in an output burst (see [IBM OUT][pp. 69 f.])

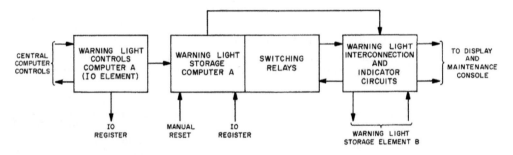

Figure 8.14: Block diagram of the warning light system (see [IBM DSP 2][p. 94])

8.4 Alarms and warning lights

Although an AN/FSQ-7/8 installation featured an intricate display system capable of graphical and textual output of track and target data, additional means were required to notify operators of conditions requiring manual intervention. These means were audible *alarms* and *warning lights* installed in the various consoles.[31] Alarms and warnings lights were controlled by the *warning light system* consisting of the *warning light control element* and the *warning light storage element*. Figure 8.14 shows a simplified block diagram of the warning light system as seen by one of the two AN/FSQ-7 computers.

The warning light control element is connected to the central computer through its IO register and is capable of receiving up to 256 bits from the computer. Therefore the warning light storage element contains 256 flip-flops arranged in an eight times 32 bits matrix. Each of these flip-flops controls up to three warning lights in different consoles or an audible alarm through relays driven by thyratron relay drivers.[32] Which of the two warning light systems, one for each computer in a duplex installation, is in control of the consoles, is determined by the associated computer's status.

---

[31] See [IBM DSP 2][pp. 92 ff.].
[32] See section 5.5.

8.5 Tape drives and card machines 163

Figure 8.15: Magnetic tape adapter (see [IBM CCS I][p. 17])

Figure 8.16: Magnetic tape drive (see [IBM CCS I][p. 18])

8.5 Tape drives and card machines

No large scale computer of the 1950s would have been complete without a complement of punch card equipment and tape drives. The AN/FSQ-7 and -8 used off-the-shelf IBM card devices[33] and tape drives[34] attached to the IO system of the computer. Data was transferred to and from these units under control of the selection and IO element.[35]

The magnetic tape subsystem consists of the *magnetic tape adapter* (shown in figure 8.15), a *tape power supply unit*,[36] and six standard tape drives type IBM 727 (shown in figure 8.16). These seven track tape drives used standard 1/2 inch tape reels at a density of 200 characters per inch. One such character consists of six bits of data and one parity bit. At a character rate of 50 μs per character, one AN/FSQ-7/8 data word of 32 bits plus one parity bit could be written every 300 μs, yielding a word transfer rate of about 3,333 words per second. An arbitrary number of words can be written

---

[33]See [IBM CCS I][pp. 221 ff.].

[34]See [IBM CCS I][pp. 227 f.].

[35]See section 6.3.

[36]This extraneous power supply was necessary since the standard tape drives required some so-called *nonstandard voltages* not available in an AN/FSQ-7/8 installation.

Figure 8.17: IBM 713 card reader with card hopper on the left (see [IBM CCS I][p. 21])

Figure 8.18: Typical IBM 723 card punch (see [IBM CCS I][p. 22])

Figure 8.19: IBM 718 line printer with exposed patch panel (see [IBM CCS I][p. 22])

sequentially to the tape as a so-called *record*. Records are delimited by an *interrecord gap* allowing the tape controller to determine the start or end of a record to be read or written.

The punch card equipment consisted of an IBM 713 card reader, an IBM 723 card punch and an IBM 718 line printer. The punch cards were divided into three groups of columns for storing binary data: The first 16 columns were not used, while the columns 17–48 and 49–80 could each hold one 32 bit machine word per row. Accordingly, a single standard 12 row punch card can hold 24 AN/FSQ-7/8 words.

9 Display system

Today, a display system would qualify as just another output system of a computer. The display systems of AN/FSQ-7/8 however were so intricate and ahead of their time that this chapter, instead of a mere section, is devoted to these. A problem like that of air defense obviously requires some way to display incoming live data in graphical form to the operators in an air defense sector.[1] Accordingly, high resolution vector displays and additional displays for textual data were developed for AN/FSQ-7/8 which are described in more detail in the following.

Basically, the display system consisted of two types of displays (apart from a projection unit used in the command post): The so-called *situation displays*, based on 19 inch display tubes with long persistence P14 phosphor,[2] capable of displaying text as well as vector images, and *digital displays*, using a special type of storage tube called *Typotron*. Both types of display were driven by respective *display generators*, one *situation display generator*, and one *digital display generator*. These display generators were attached to the central computer through the magnetic drum system. The situation display generator was controlled by the six and nine fields of the TD and RD drums, while the digital display generator was controlled by the DD field of the MIXD drum:

> "The processing ability of the buffer devices is fully exploited in the display system. In this case, the central computer maintains a coded table on the buffer display drum. This table is interpreted and displayed by special-purpose equipment every $2\frac{1}{2}$ seconds at the appropriate console. The central computer can change any part of the display at any time by rewriting only appropriate words on the drum."[3]

Since these displays were the main source of information for the operating personnel at a DC or CC, usability was a prime concern during their development:

> "A great deal of engineering went into the display system electronics and the logic of what to display and what operator actions were required at each operator position. There was also the question of console layout for ease of operation. With a characteristic direct approach, one of the engineers ran a simple experiment. He suspended a Charactron display-tube shipping carton from the ceiling

---

[1] Colloquially, operators on the display consoles were known as *"scope dopes"*.
[2] This coating emits a purple flash of light when hit by an electron beam, followed by an orange afterglow persisting for several seconds. It should be noted that a second type of SD console used to create photographic records of current air situations employed display tubes with P11 phosphor. This phosphor produces a very bright blue flash of light, well suited for photographic purposes with a negligible afterglow. See [DYALL 1948] for more information about early research at MIT on display phosphors.
[3] See [EVERETT et al. 1983][p. 338].

of his office by means of four strings and drew the outline of the tube surface on one end of the carton. Visitors were seated before the mock-up and asked to adjust the strings to what they thought was the best height and tilt angle of the mocked-up tube face. Height of the front and read of the carton were then measured and recorded. The measurements from many trials were averaged, and the geometry of the Cape Cod displays was determined."[4]

While today's display systems employ high refresh rates at 70 Hz and above to create a flicker-free picture, this was out of the question given the available computing power of the AN/FSQ-7/8 and the fact that a typical installation contained more than 100 displays. Thus other techniques had to be employed in the display systems to ensure steady pictures. While the situation displays used a special long persistence phosphor, the digital displays were based on an analog storage tube quite similar to tubes in storage oscilloscopes of those days. Both techniques had a severe drawback: The pictures displayed were very dim and the situation displays grew dimmer every second between two refresh cycles of the central computer. Accordingly, the areas where displays were located required special low intensity non-glare indirect lighting using blue light sources.[5] These light sources were mounted on the ceiling of the display rooms atop a honeycomb structure which eliminated any stray light which might cause glaring, thereby considerably making the overall impression even dimmer which raised some initial concerns regarding the practicability of the overall display system:

"The broad-band blue operational lighting system designed for the SAGE operations rooms was shown to WE-ADES and Air Force building personnel on 24 May 1955. It was their impression that the blue environment caused dizziness and vertigo upon entering the room. There was no agreement among them as to whether the effects tend to disappear with time or not. At a subsequent meeting on 1 June 1955 in New York a draft specification of the system was discussed. Since we were unable to resolve the vertigo problem in New York, still another meeting has been scheduled here at Lincoln for 9 June 1955. At this time the Air Force building people will be accompanied by several experts of their choice who will help them evaluate the effects of blue light.

Since these meetings, we have contacted many people with medical, psychological, and operational backgrounds, and we can find no one who knows of any physical problems that arise from the blue environment. Some people did state that there may be mental reasons for feeling depressed under the blue lights. However, this problem is not severe enough to reject the system when compared to the difficult operating conditions existing in the extremely low levels of light that would be necessary if we did not restrict the ambient light to the blue region.

The operational people were completely unanimous in their approval of the system, because it allowed them to operate at light levels much beyond anything they had previously experienced and, furthermore, it does not require dark adaptation upon entering, nor the momentary blindness on leaving the room."[6]

---

[4]See [WIESER 1983][pp. 367 f.].
[5]Accordingly the display rooms were known as *blue rooms*.
[6]See [Biweekly Report 3674][p. 15].

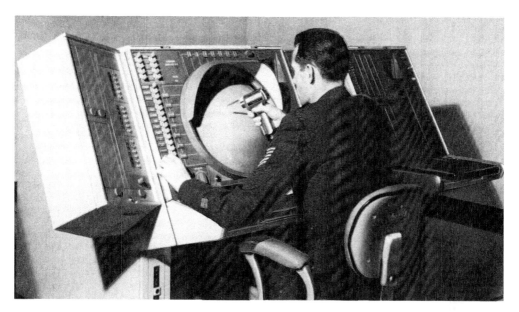

Figure 9.1: Situation display with light gun (see [IBM DSP 1][p. 19])

The display consoles were essentially grouped into four groups: *Surveillance, identification, crosstell,* and *intercept control.* The surveillance consoles were basically digital radar displays similar to PPI displays on which the last seven scans were always shown, thus creating trails which could be assigned track numbers using the light gun.[7] Newly assigned tracks were then routed to the identification section. The task of the operators was to determine whether an unknown track was friendly or hostile. Therefore they needed access to all flight plans of civil and military organizations against which these tracks were matched. Operators at crosstell display consoles[8] managed tracks crossing adjacent sectors while those at the intercept control consoles coordinated the interceptors in case of hostile tracks.

9.1 Situation display

Figure 9.1 shows a typical situation display with its prominent circular 19 inch display tube in the middle and a so-called *special side wing.*[9] The operator holds a light gun pointing at some object of interest displayed on the screen. The upper right of the central console frame holds the small DD tube, mostly hidden behind the light shade of the SD tube. To the right of the console an *auxiliary display console* with an additional DD tube mounted in the upper right corner can be seen.

---

[7] See section 4.2.
[8] "[The] name [was] inherited from an earlier manual system in which human operators followed the tracks of aircraft on radar screens and coordinated matters by talking to one another on telephones.", see [ORNSTEIN 2002][p. 23].
[9] See [IBM DSP 1][p. 15].

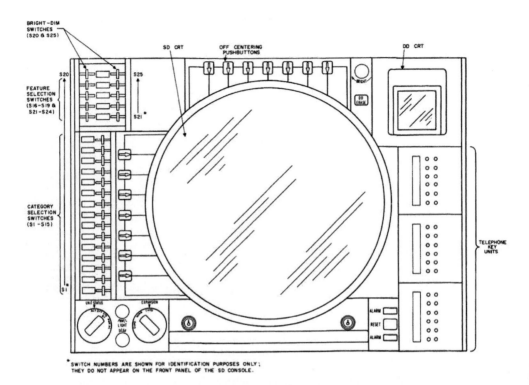

Figure 9.2: Situation display layout (see [IBM DSP 1][p. 114])

Figure 9.2 shows the layout of the situation display console: The *category selection* and *feature selection switches* on the far left were used to control which type of messages would be displayed at this particular console. A typical air situation resulted in too many different message types to be displayed at once on a display – therefore the operator could select certain message types which were of interest to him. Using the *bright-dim switches* the brightness of displayed feature groups could be selected.

The display itself was divided into seven horizontal and seven vertical areas. Using the *off centering push-buttons* located on the left and on top of the SD tube, it was possible to select a particular screen area for an expanded view under control of the rotary *expansion switch* on the lower left. This feature was the result of experiences during early experiments with Whirlwind:

> "The need for a X8 expansion for the Weapons Director (WD) console is indicated by the large number of reports related to the symbology overprinting problem. [...] 16 incidents were related to this problem and all recommendations called for the provision of X8 as a solution.
>
> Heavy overprinting causes many problems for the WD in reading track symbology and results in considerable delay in taking appropriate console actions."[10]

---

[10]See [ASTIA 401 412][p. 25].

9.1 Situation display

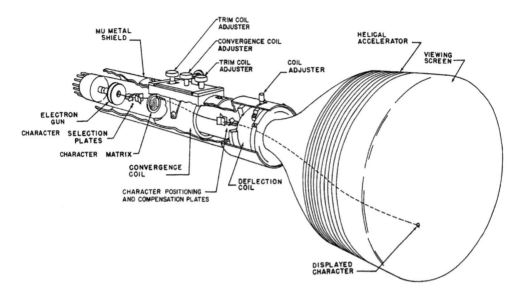

Figure 9.3: Principle of operation of the 19 inch SD display tube (see [IBM DSP 1][p. 48])

The situation display consoles were intricate and delicate devices and very sensitive to over-temperature as [CEM 55-19][Atch. 11] makes clear: *"Display consoles will overheat in two minutes if air flow is stopped."*

Figure 9.3 shows the principle of operation of the special display tubes used in the SD consoles:[11] A rather wide electron beam generated by the electron gun on the far left is deflected by a first set of deflection plates called *selection plates*. The electron beam controlled by these plates passes a character forming matrix at a certain position. This matrix, a thin etched steel plate effectively acting as a stencil, contains the shapes of all 64 displayable characters[12] and acts as shown in figure 9.4. After passing through, the electron beam runs through a

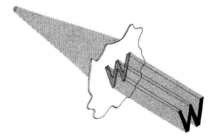

Figure 9.4: Stencil effect of the character matrix (see [IBM DSP 1][p. 45])

convergence coil and a second pair of deflection plates. These plates, mounted in reverse, recenter the shaped electron beam before it enters a deflection coil which finally determines the display location on the screen.

As complicated as this scheme of character generation looks from today's point of view, it was the only viable way back then to generate high-resolution characters on a multitude of display tubes without requiring too much computer time, thus simplifying the overall display system considerably. Figure 9.5 shows a typical output generated

---

[11] See [IBM DSP 1][pp. 37 ff.].
[12] Only 63 of these characters were actually etched into the stencil. The 64th blanked the electron beam, thus allowing the display of a blank in a message (see [IBM DSP 1][p. 44]).

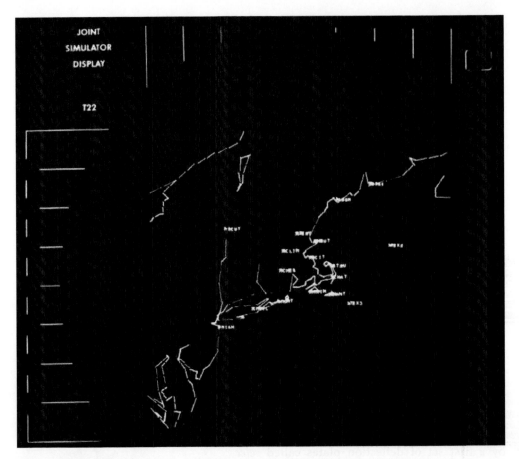

Figure 9.5: Typical situation display showing the coastline of New England and adjacent installations (see [EVERETT et al. 1983][p. 334])

by a situation display. To generate vectors for displaying tracks, coast lines etc., the electron beam is focused before entering the character matrix where it is deflected to a rectangular area larger than its diameter. This beam is then deflected by the magnetic deflection system to create the desired vector pattern on the screen.

The so-called *situation display generator element*[13] received data from the TD and RD drums and generated the necessary control signals for all situation displays in a DC or CC. The TD drum was capable of holding up to 1,536 messages of 256 bits each for the SDGE, while the RD drum could store a maximum of 16,384 24 bit messages of radar data for display.[14] Based on these messages the SDGE generated a variety of signals controlling the SDs – the most important of these are the position of a message to be displayed, transmitted by two digital signals XL and YL, while character positioning and character selection were controlled by analog signal pairs (SD $\pm X, Y$) and

---

[13] *SDGE* for short.
[14] See [IBM DSP 1][p. 151].

9.2 Light gun 171

Figure 9.6: Situation display generator (see [IBM DSP 1][p. 6])

(SD X, Y). In addition to controlling the SDs, the SDGE also transmitted target data back to the MDI unit where it was combined with trigger data received from light guns. Figure 9.6 shows the SDGE.

9.2 Light gun

As if high-resolution graphical displays would not have been impressive enough in the 1950s, they also gave birth to an innovative input element, the *light gun*.[15] Although an operator would use the various push-buttons on the side wings of the SD console to setup messages for the central computer denoting actions to be taken or information being requested, the light gun was necessary to associate these manually entered messages with a particular track or target display on the SD.[16]

---

[15] The name alone shows the military background of this device – its commercial counterpart would become known as *light pen* in years to come.

[16] Pulling the trigger of a light gun released a message for further processing by the active computer just as pressing the activate push-button would have done (see section 8.2).

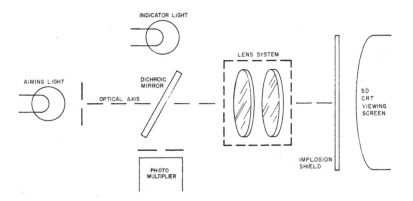

Figure 9.7: Optical system of a light gun (see [IBM DSP 2][p. 136])

Figure 9.7 shows the optical path of a light gun: At its heart is a photomultiplier tube sensitive to the blue streak of light emitted by the phosphor of the SD tubes when hit by the electron beam. A dichroic mirror is used to deflect light from the surface of the SD tube to the photomultiplier while simultaneously shining an aiming light onto the screen. The indicator light is activated when a target has been successfully captured.

Figure 9.8: Detail view of a light gun (©Marcin Wichary)

Figure 9.8 shows a light gun – clearly visible are the trigger button on the grip and the protruding pointer which serves as an aide for pointing at a particular object. Additionally, this pointer assures that the light gun is used at the correct distance as determined by the focal length of its optical system.

The output signals from the photomultiplier tube were amplified in a three-stage amplifier housed in a dedicated *light gun amplifier*. The output of this amplifier, the trigger signal generated by the trigger pushbutton, and a computer generated *pass light gun gate* signal are then fed into a three-input AND which in turn controls a thyratron relay driver switching from the aiming light to the indicator light and generating a light gun pulse for the MDI unit.[17]

The activated relay short circuits the thyratron tube via a dedicated contact pair and the still depressed manual trigger, thereby automatically resetting the thyratron.

---

[17] See [IBM DSP 2][p. 135].

Figure 9.9: Pickup of an area discriminator (see [IBM DSP 2][p. 138])

9.3 Area discriminators

During peacetime, friendly tracks, such as generated by registered civil and military flights, are much more common than hostile ones. If each such friendly track had to be assigned manually by an operator using a light gun, an extremely high and unnecessary workload would have been placed on the operators. To automate the detection of known friendly tracks, so-called *area discriminators* were used.

These area discriminators worked similarly to the GFI mapper consoles[18] but were based on modified SD consoles as shown in figure 9.10. Two such systems were used in a DC to detect friendly tracks.[19] The actual area discriminator – basically the optical pickup system of a light gun without aim and indicator lights and without the manual trigger push-button (shown in figure 9.9) – is mounted on a frame in front of a SD console. The photomultiplier of the pickup unit is sensitive only to the blue streak of light emitted by the phosphor coating of the display tube when excited by the electron beam, but is insensitive to the orange afterglow.

Figure 9.10: Area discriminator console (see [IBM DSP 2][p. 126])

Using a special grease pencil containing a masking fluid opaque to blue light but transparent for the orange afterglow of the SD display tube, all areas on the screen belonging to known friendly tracks were manually masked. This was done based on incoming *ATC* flight plan data.

---

[18]See section 8.1.2.
[19]See [IBM DSP 2][p. 126].

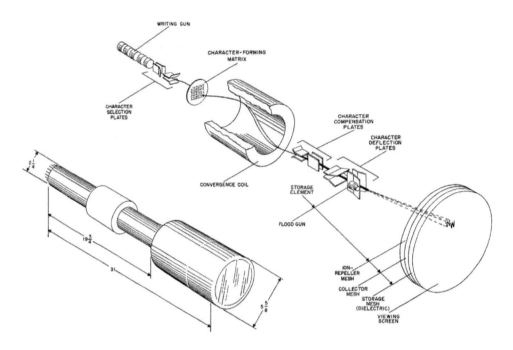

Figure 9.11: Principle of operation of the Typotron display tube (see [IBM DSP 1][p. 57])

9.4 Digital display

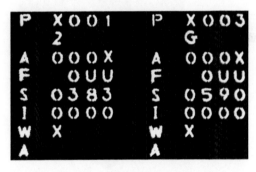

Figure 9.12: Typotron display example

The large 19 inch display tubes of the SD consoles were accompanied by smaller five inch DD tubes capable of text output only. Character generation in this tube works similar to that in the 19 inch SD tubes as shown in figure 9.11:[20] A slightly defocused electron beam is passed through a character forming matrix, controlled by a pair of deflection plates. After passing a static convergence coil, the shaped electron beam is realigned with the tube's axis by means of a pair of *character compensation plates*. The position of a character to be displayed on the screen is then determined electrostatically by a pair of deflection plates.[21] Figure 9.12 shows a screenshot from a typical Typotron display in a SAGE installation. Clearly visible are the small obstructions caused by the character forming matrix in the "0" and "P" characters.

---

[20]This tube was actually developed by HUGHES Aircraft Corporation and was marketed under the label *"Typotron"*. See [HUGHES Aircraft Corporation] for more information.
[21]The SD display tubes use a magnetic deflection system.

9.4 Digital display

Figure 9.13: Digital display generator element (see [IBM DSP 1][p. 78])

While the large 19 inch SD display tubes rely on long persistence P14 phosphor, the DD tubes employ P1 phosphor which emits a brilliant green spot of light with short to medium persistence when hit by an electron beam. Therefore another means for storing information on the screen had to be used. The heart of this storage system is the *flood gun* shown in figure 9.11 in conjunction with the *ion repeller, collector,* and *storage mesh* located behind the screen. The characters to be displayed are stored as charge patterns on the storage mesh similar to the technique used in the MIT storage tube described in section 3.2.1. The flood gun produces a wide beam of low-velocity electrons. These can pass the storage grid only on places where a positive charge has been created due to secondary emission by displaying a character. Although these charge patterns will eventually degenerate over some minutes, using this technique it is possible to store a message generated once for a prolonged time. Accordingly, burn-in patterns on the phosphor coating of the screen resulted when no update data was sent for a prolonged time. In contrast to the large situation displays which were updated automatically, the digital displays remained static unless an explicit update was sent from the active computer.

All DD tubes were controlled by the so-called *Digital Display Generator Element*, DDGE for short, shown in figure 9.13. This unit was coupled to the active computer via the DD fields of the MIXD drum.

Figure 9.14: SAGE CC-01 Subsector Command Post

9.5 Photographic recorder-reproducer element

The SD and DD consoles were not suitable for use in the *command post* of a DC or CC like that shown in figure 9.14, where several persons needed an unobstructed view of the current air situation at once. Such a large scale picture was generated by a KELVIN-HUGHES-like projector, called *Photographic Recorder-Reproducer Element, PRRE* for short, which aptly describes its basic principle of operation: A 35 mm film was exposed to the picture displayed by a seven inch display tube similar to the 19 inch SD tube.[22] This film then entered a rapid film processor developing and fixating the image by subjecting it to sprayed chemicals. The six inch display area of the display tube was optically scaled down to a 0.8 inch circle on the film.[23] The next station was the actual projector where the image on the film was displayed onto a large 14-by-14 foot screen in the command post through a lens system and a high intensity mercury vapor lamp, see figure 9.16.

The film was always moved one frame at a time, so that one picture was recorded, its predecessor was being developed, and the picture before that was being displayed. The overall process from exposure to projection took only 30 seconds.[24] Figure 9.16 shows the PRRE unit while figure 9.17 shows a typical screenshot from a command post display.[25]

---

[22] This tube used a short persistence P11 phosphor emitting bright blue light, better suited for photographic purposes than the P14 phosphor used in the SD consoles.

[23] See [IBM DSP 2][p. 32].

[24] See [IBM DSP 2][p. 40].

[25] Picture source: http://upload.wikimedia.org/wikipedia/commons/f/ff/SAGE_control_room.png, retrieved 12/14/2013.

9.5 Photographic recorder-reproducer element

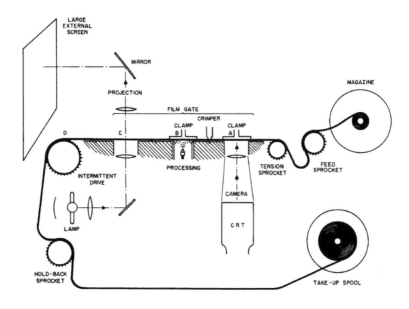

Figure 9.15: Principle of operation of the PRRE (see [IBM DSP 2][p. 34])

Figure 9.16: PRRE unit (see [IBM DSP 1][p. 11])

Figure 9.17: Picture displayed by projector[26]

---

[26]See http://en.wikipedia.org/wiki/File:SAGE_control_room.png, retrieved 04/14/2014.

10 Machine consoles

Figure 10.1: Maintenance control console room of an AN/FSQ-7 installation (courtesy Air Force History Support Office)

A giant duplex computer installation like an AN/FSQ-7 or -8 requires equally giant maintenance operation consoles to monitor the computers and their associated input/output equipment. These consoles were icons not only of their time but are still used today in a variety of movies whenever the impression of a giant computer is required. So even if there is no longer any AN/FSQ-7 still operating, at least parts of its maintenance consoles are still used to represent the archetypal electron brain.

Figure 10.1 shows a view of the maintenance control console room of an AN/FSQ-7 installation looking from the *simplex maintenance console* (not visible here) across the room. On the left and right are the two *duplex maintenance consoles*, one for each of the two computers. In the center is the IBM 718 line printer while the *duplex switching console* can be seen in the middle of the far end of the room.

An early rendition of a duplex maintenance console is shown in figure 10.2. It contains a plethora of neon indicators giving an insight into the current operation of one of the two duplexed computers and its attached input/output systems. It was quite common to place a Polaroid camera mounted on a tripod in front of this console when

Figure 10.2: Duplex maintenance console with display controls (see [IBM DSP 1][p. 3])

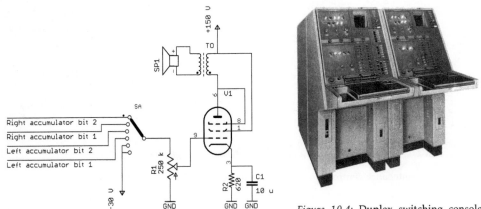

Figure 10.3: Console audio amplifier (see [IBM CCS XD][p. 412])

Figure 10.4: Duplex switching console (see [IBM CCS I][p. 23]

a computer had crashed to take a snapshot of the register contents as displayed by the hundreds and hundreds of neon lights. These pictures could later be analyzed to determine the cause of the crash. In addition to this, the duplex consoles featured a speaker with an associated audio amplifier as shown in figure 10.3: By means of a selector switch, this amplifier could be connected to one of the two leftmost bits of the left or right arithmetic unit of the computer. An experienced operator could recognize the patterns of humming noises emanating from the speaker. Using this feature, endless loops, sudden changes in the operation of the computer etc. could be spotted in an instant.

Typically, one of the duplexed computers was the *active* computer while the second system was in standby mode. Under normal conditions, this second computer monitored the active computer and periodically copied live data from this to minimize switchover time in case of an emergency like a crash or hardware fault on the active

10 Machine consoles

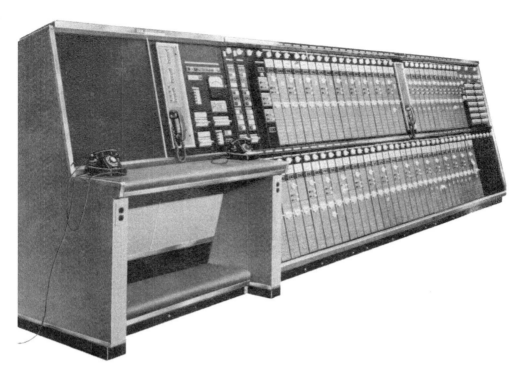

Figure 10.5: Simplex maintenance console (see [IBM DSP 1][p. 9])

side.[1] Under these ideal circumstances, a switchover was only a matter of seconds and was performed using the *duplex switching console* shown in figure 10.4. A scheduled switchover, which occurred on a daily basis, transferred approximately half of the core memory's contents from the active to the standby computer as well as most of the current data on the LOG, MIXD, RD, and TD drums.[2]

Not visible in figure 10.1 is the so-called *simplex maintenance console* which is located on the front side of the maintenance console room from where this picture was taken. This console is shown in detail in figure 10.5. It contains all control and display elements for simplex equipment in an AN/FSQ-7/8 installation.[3] From left to right, these are an optional blank panel for later addition of a maintenance oscilloscope and probe facilities, an intercommunication telephone, the marginal checking panel and the many control panels for GFI, Computer Entry Punch (CEP), LRI, XTL, and TPG.[4]

---

[1] The role of active and standby computer was altered on a daily basis to provide the necessary time slots for maintenance tasks on both machines.

[2] See [VANCE et al. 1957] for more information about duplex switching etc.

[3] *"Generally, equipment is simplex if its failure will not seriously affect the defense functions of the Direction Central."* See [IBM INPUT][p. 228].

[4] See [IBM INPUT][pp. 16 f.].

11 Power supply

Powering a DC with its duplex AN/FSQ-7, the complex input/output equipment, and its more than 100 display consoles is a formidable task.[1] Such an installation required roughly 3,000 kW of which 1,000 kW were used for the duplex computer system with all of its associated equipment while the remaining 2,000 kW were required for lighting and the air conditioning system. Since power outages often had devastating effects on the vacuum tubes in the computer and its associated equipment,[2] utmost stability and availability of the power supply equipment were of prime importance.

11.1 The powerhouse

Each DC and CC featured a dedicated *powerhouse* which can be seen in figure 4.5 in section 4.3 as the flat building directly next to the four story DC blockhouse. This building housed five diesel engines with attached generators, each capable of delivering 650 kW as 480 V AC power. Figure 11.1 shows one of these diesel generator units installed at DC-03 at Syracuse in 1956.

Figure 11.1: Diesel engines and generators at DC-03, Syracuse, September 1956 (courtesy of Air Force Historical Research Agency)

The powerhouse also contained two so-called *load centers*, each featuring a 300 kVA transformer yielding 120 V and 208 V from the 480 V supply and associated circuit breakers, and three *substations* containing one 500 kVA transformer each. Only two substations were actually necessary for operation of the DC or CC, the third could replace a failed substation or load center in case of an emergency.[3]

Figure 11.2 gives an impression of the overall power distribution system. On top are the diesel generators with the associated load centers and substations while the equipment shown in the lower half was located in the DC building itself. Although the powerhouse had its own maintenance and operations staff, they were closely knit with

---

[1] See [IBM POWER] for a thorough description of the overall power supply system of an AN/FSQ-7/8 site.
[2] *"We had two power failures last July [1979] and lost about 400 tubes."*, see [SKYWATCH][p. 7].
[3] See [IBM POWER][p. 3].

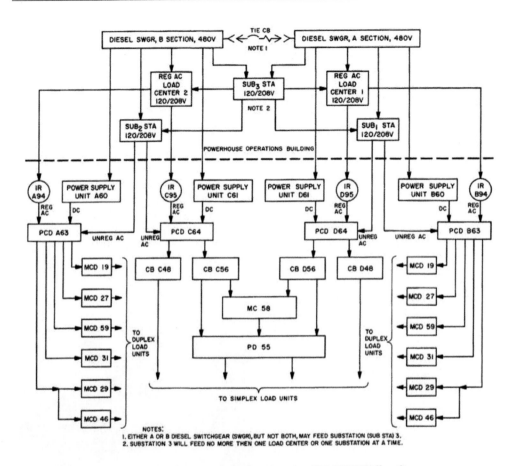

Figure 11.2: Load centers and substations (see [IBM POWER][p. 4])

their counterparts in the blockhouse operating and maintaining the computer and air conditioning.

11.2 Regulated power supplies

The raw output voltages generated by diesel driven generators are, of course, not suitable for running an intricate and fragile digital computer like the AN/FSQ-7/8. This task required a wide variety of regulated AC and DC voltages at high currents.

AC regulation was done by *induction regulators* fed by the 480 V output from the powerhouse. These regulators were basically motor-driven *variable induction transformers*.[4] These AC regulators typically delivered three phase power at 120 V and 208 V to loads such as the filament transformers in the computer.

---

[4]See [IBM POWER][pp. 13 ff.] for more information on these devices.

11.2 Regulated power supplies

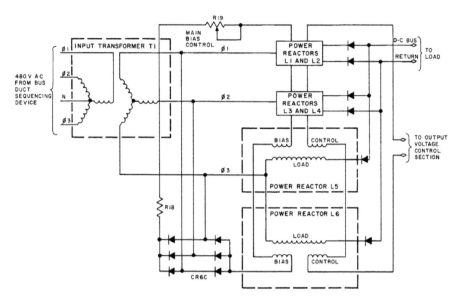

Figure 11.3: Principle of operation of a typical DC power supply (see [IBM POWER][p. 31])

More interesting are the regulated DC power supplies: To generate the multitude of different stabilized DC voltages required by the duplexed computers and their simplex equipment, the 480 V AC output delivered by the powerhouse was transformed to a voltage near the desired DC level in a first step, by means of a suitable step-down transformer as shown on the left of figure 11.3. The secondary of this transformer then fed a number of *power reactors*. These are essentially *saturable reactors*, i. e. transformers with additional control windings allowing to change the saturation of the iron core. The power reactors shown in figure 11.3 feature two such control windings, one *bias* winding and one *control* winding. Coarse setting of the DC output voltage was done by adjusting the bias current while a regulation circuit generated the necessary current to drive the control windings.

The control circuit driving the inputs denoted *"to output voltage control section"* was based on a *magnetic amplifier* – basically just another saturable reactor. The saturation control winding of this amplifier was driven by a current derived from the voltage difference between the power supply's DC output and a reference voltage derived from a ZENER diode[5] circuit. Small changes in this difference had large effects on the inductive resistance of this magnetic amplifier and caused a corrective current driving the control windings of the power reactors.[6] Table 11.1 shows the various stabilized DC voltages required for an AN/FSQ-7 installation with typical current ratings.

---

[5]The so-called ZENER diode is named after CLARENCE MELVIN ZENER, 12/01/1905–07/15/1993, and exhibits a rather stable and well-controllable breakdown voltage when driven in reverse direction which can be used to generate a voltage reference.

[6]Obviously, such a magnetic amplifier exhibits a time constant depending on the frequency of the signal driving its primary. To minimize the reaction time of the control circuit, a 1,440 Hz signal was used to drive the magnetic amplifiers. This voltage was generated by a motor-generator unit, called a *high-frequency alternator*, feeding all magnetic amplifiers in the various DC power supplies.

	Rated current (A)			Rated current (A)	
Voltage	Simplex	Duplex	Voltage	Simplex	Duplex
+600	85	3	−30	70	120
+250	50	65	−150	80	225
+150	75	120	−300	30	55
+90	60	130	−48	560	450
+10	100	45	−72		13
−15	10	20			

Table 11.1: Current ratings (see [IBM POWER][p. 30])

Unregulated AC from the two operational substations, regulated AC from the load centers and induction regulators, and regulated DC were fed to four *power control and distribution* units, *PCD* for short, in each installation. Two of these units fed the duplexed computers while two supplied the simplex equipment. Fed by the PCDs are the *marginal checking and distribution* units, *MCD* for short which in turn fed the computer, input/output, and display equipment.

11.3 Power distribution

Figure 11.4 shows the overall schematic of the power distribution system of an AN/FSQ-7 installation: The four PCD units are shown at the top. The duplex equipment (the central computers, display systems, manual input system, warning light system, drums, output system etc.) were powered by twelve MCD units described in the following section.

The official procedure of applying power to a duplex computer system gives an impression of the complexity of the task as well as of the fragility of the vacuum tube equipment:[7]

> *"The following procedures are to be used as general guidelines to recycle duplex power to a computer:*
>
> a. *After a power failure or a scheduled power outage, open the AC input switches to both main and auxiliary drums.*
>
> b. *Obtain approval from power plant personnel to recycle power. Depress the AC Only push button firmly. If power fails to recycle properly, inform the power plant personnel and check our system for failure indications. Do not attempt to recycle power until the OK is received from power plant.*
>
> c. *Allow a maximum of 10 minutes to warm up the filaments prior to applying DC to the loads.*
>
> d. *Request unregulated AC be reapplied if required.*
>
> e. *Check and insure filament voltages are set to 120 volts AC.*

---

[7] See [CEM 55-19][Atch. 10].

11.3 Power distribution

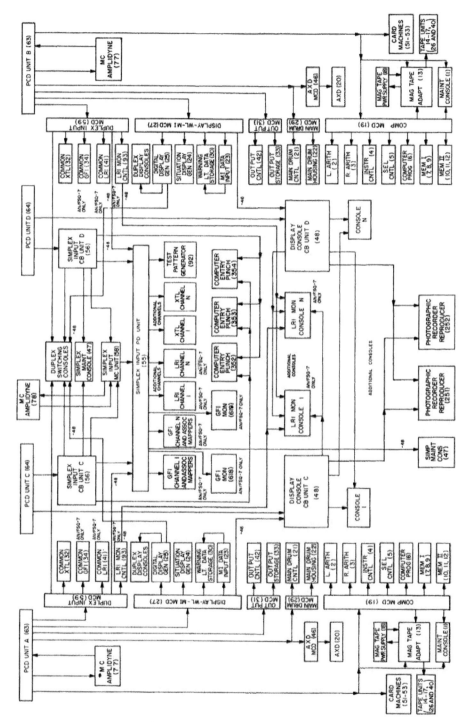

Figure 11.4: Block diagram of the power distribution system for an AN/FSQ-7 (see [IBM POWER][p. 46])

f. *After the filament warmup has elapsed, depress the power on push button. There is no requirement to notify the power plant prior to this action. All units may have DC applied at the same time via the Power On push button.*

g. *Apply power to the drum motors. Do not attempt to cycle main and auxiliary drums at the same time.*

h. *Repair any obvious failures, then begin the computer checkout with ADIOS-AUOC. Once a failure is detected, run any program that may help solving the problem. AUOC will give an overall analysis; therefore, possibly enabling two problems to be worked on simultaneously. Marginal check program passes are desirable if they can be accomplished during the troubleshooting and repairing of other areas of the computer.*

i. *Closeout the ESR after AUOC has cycled error-free for 10-15 minutes."*

11.4 Marginal checking system

Marginal checking was pioneered in Whirlwind and proved necessary to achieve the required margins of reliability of the computer systems used in the air defense system:

"Marginal checking is a preventive maintenance procedure used to increase the reliability of AN/FSQ-7 and -8 equipment. A computer containing 5,000 tubes and 10,000 diodes can be expected to fail every half hour after the circuit elements are initially aged. Since the number of diodes and tubes in AN/FSQ-7 and -8 equipment exceeds these quantities, computer failure can be expected more frequently than every half hour. The Marginal Checking System is incorporated to increase the useful operating time of the equipment by determining the components which are likely to fail between scheduled maintenance periods. In addition to determining the reliability of the equipment, the Marginal Checking System also aids maintenance personnel in diagnosing and locating troubles."[8]

Basically, marginal checking may vary filament voltages and stabilized DC voltages of a system. Filament marginal checking was only used in some display consoles, an application where reduced emissions result in a dimmer picture. The duplex computer systems with all of their associated equipment were subjected only to variations of their most critical DC supply voltages.

Figure 11.5 shows a typical equipment *life curve*: The x-axis denotes time of operation of equipment while the y-axis shows the allowable margin[9] of variation concerning a vital parameter such as supply voltage. The voltage at which a particular circuit still operates reliably is determined by deliberately varying its supply voltages. Based on this voltage, the operational margin of the equipment under test is determined.

---

[8]See [IBM MC][p. 9].

[9]*"The amount of variation from the nominal value that can be introduced before circuit failure occurs is called the margin of reliability of that circuit"*, see [IBM MC][p. 9].

11.4 Marginal checking system

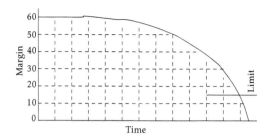

Figure 11.5: Typical life curve

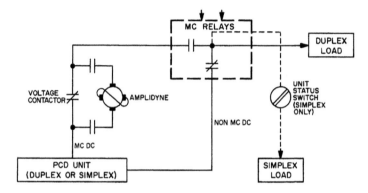

Figure 11.6: Principle of operation of marginal checking (see [IBM MC][p. 13])

This margin is then compared to a predefined limit – if it is less than this limit, the circuit will be replaced instead of waiting for it to fail due to aging processes. Such life curves were determined for all of the equipment being subject to marginal checking. To simplify fault isolation, marginal checking was applied to subsystems as small as possible at once.

The +250 V, +150 V, +90 V, −150 V, and −300 V supply lines could be varied by means of a so-called *amplidyne*. An amplidyne is basically a motor-generator unit with an externally driven field-control winding on the generator side. Changing the current flowing through this winding changes the output voltage of the generator. Each of the twelve marginal checking systems contained one amplidyne, so only one of the aforementioned voltages could be varied in an equipment group at a time. These amplidynes consisted of a 3 kW 125 V DC generator and a 208 V, 3-phase, 5 hp motor running at 1,800 rpm. The gain of such an amplidyne configuration is remarkable – typical values are in the range of 10,000 to 50,000, so small changes in the field-control winding result in rather large changes at the generator's output. As shown in figure 11.6, the amplidyne was switched in series to the voltage under excursion during a marginal check operation.

Marginal checking was performed only on equipment in the standby state as determined by switches at the respective control consoles. The MC system allowed three modes of operation: *Calculator mode* in which the MC system was under program con-

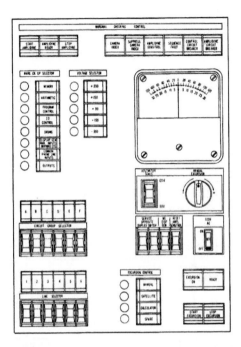

Figure 11.7: MC control panel, duplex maintenance console (see [IBM MC][p. 44])

trol,[10] *manual mode*, which was controlled from the MC control panel located in the maintenance console room and was normally used to perform a drill-down analysis to isolate units at fault which could not be identified during a calculator mode run, and, finally, *satellite mode*. This mode was similar to manual mode with the exception that a portable potentiometer was used to control the voltage excursions. This mode allowed marginal checking under manual control while servicing the computer system.[11]

Figure 11.7 shows the MC control panel located on the duplex maintenance console. The switches on the left select the unit (memory, arithmetic etc.), the circuit, supply line, and voltage to be varied.[12] The right hand side contains a voltmeter indicating the amplidyne output voltage and controls for manual excursions.

---

[10] A couple of operate- (PER) and branch-instructions were reserved for the control of the MC-system: *Start Excursion* (**PER 21**), *Stop Duplex MC Excursion* (**PER 22**), *Stop Simplex MC Excursion* (**PER 23**), *Sense Duplex MC Excursion On* (**BSN 20**), *Sense Simplex MC Excursion On* (**BSN 27**), and **HLT** (see [IBM MC][pp. 31 ff.]).

[11] This satellite mode was not implemented on all installations.

[12] These switches were operative only in manual or satellite mode (cf. [IBM MC][p. 45]).

12 Programming

Programming an early computer like the AN/FSQ-7 was quite different from today's perspective as many of the modern programming idioms, such as using a stack for parameter passing to subroutines and storing return addresses etc. were not used back then. In addition to that, the two 16 bit arithmetic elements of the AN/FSQ-7, working in parallel, are still a rather unusual feature.

The following sections give an overview of the instruction format, instruction set, and typical programming idioms with regard to indexing and subroutine handling. The final AN/FSQ-7/8 computers featured 59 different instructions grouped into eight classes, while the XD-1 prototype implemented only 48 instructions. All following examples and tables are based on this original set of instructions.

12.1 Instruction format

The instruction set of AN/FSQ-7/8 looks quite *RISC*[1]-like from a modern point of view: Each instruction occupies one machine word of 32 bits (plus one parity bit) and there are no provisions for intricate operand address manipulations apart from a simple index adder. Nevertheless, the instruction format shown in figure 12.1 looks a bit cluttered at first sight.

The right half-word of each instruction contains its (in many cases optional) operand address while the left half-word specifies the instruction to be executed. This straightforward addressing scheme was sufficient for the early incarnation of AN/FSQ-7 with its two core memory systems of only 4 k words each. With the replacement of one of these memories by a 64 k system, a total of 68 k words of main memory (plus 32 addresses for the test memory) became available, rendering the 16 bits of the right half-word insufficient to store an address. Therefore the sign bit of the left half-word – prior to this unused in the instructions – was logically assigned to the address portion of an instruction, thus allowing a 17 bit address space, more than sufficient for the 4 k plus 64 k complement of main memory. When addressing memory, the sign bits lose their special meaning since addresses are always unsigned quantities.

The first three bits of an instruction specify which – if any – of the available five index registers is to be used for generating an operand address.[2] If no indexed addressing is to be used, these bits contain 000. The next three bits select the instruction class. The remaining bits specify the variation of the instruction and a lot of other details depending on the instruction class and variation.

---
[1] Short for *Reduced Instruction Set Computer*.
[2] Of course, this information has only an effect on instructions supporting indexed addressing at all.

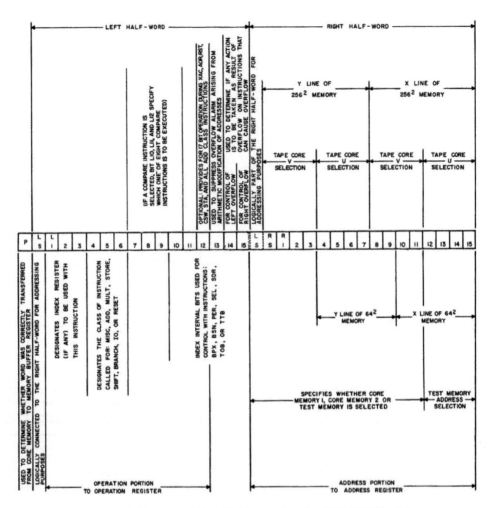

Figure 12.1: Structure of AN/FSQ-7 instructions (see [IBM CCS I][p. 94])

12.2 Instruction set

Table 12.1 shows all 48 instructions present in the XD-1 and XD-2 systems in 1955, being clustered into eight instruction groups, each containing up to sixteen instructions. These instructions will be described in more detail in the following subsections[3] using a C-like notation to describe the operation of the instructions. So mem[address + index] denotes the contents of the memory location addressed by an operand address plus the contents of an index register etc. The abbreviation AC for the accumulator always denotes the left and right accumulator of the two arithmetic elements. The same holds true for the A and B registers. ":=" denotes the assignment operation.

---

[3]See [IBM PGM][pp. 396 ff.]. Appendix B contains some programming cards for AN/FSQ-7 as they were used during programming.

12.2 Instruction set

Class	Binary code	Variation	Indexable	Binary code	Mnemonic
Miscellaneous	000	Program Stop		0000	HLT
		Extract	✓	0001	ETR
		Operate		001-	PER
		Clear & Subtract Word Counter		0100	CSW
		Shift Left & Round		0101	SLR
		Load B Registers	✓	0110	LDB
Add	001	Clear & Add	✓	0000	CAD
		Add	✓	0001	ADD
		Twin & Add	✓	0010	TAD
		Add B Registers to Accumulator Registers	✓	0011	ADB
		Clear & Subtract	✓	0110	CSU
		Subtract	✓	0111	SUB
		Twin & Subtract	✓	1000	TSU
		Clear & Add Magnitude	✓	1100	CAM
		Difference Magnitude	✓	1101	DIM
Multiply	010	Multiply	✓	1010	MUL
		Twin & Multiply	✓	1011	TMU
		Divide	✓	1100	DVD
		Twin & Divide	✓	1101	TDV
Store	011	Store	✓	0101	FST
		Left Store	✓	0110	LST
		Right Store	✓	0111	RST
		Store Address	✓	1000	STA
		Right Add 1	✓	1001	AOR
		Exchange	✓	1010	ECH
		Deposit	✓	1100	DEP
Shift	100	Shift Left		0000	DSL
		Shift Right		0001	DSR
		Shift Accumulators Left		0100	ASL
		Shift Accumulators Right		0101	ASR
		Left Element Shift Right		1000	LSR
		Right Element Shift Right		1001	RSR
		Cycle Left		1100	DCL
		Cycle Accumulators Left		1110	FCL
Branch	101	Branch on Positive Index[4]		001-	BPX
		Sense		010-	BSN
		Branch on Zero		1000	BFZ
		Branch on Minus		1001	BFM
		Branch on Left Minus		1010	BLM
		Branch on Right Minus		1011	BRM
Input-Output	110	Load IO Address Counter	✓	0000	LDC
		Select Drum	✓	001-	SDR
		Select	✓	010-	SEL
		Read	✓	1110	RDS
		Write	✓	1111	WRT
Reset	111	Reset Index Register		1011	XIN
		Reset Index Register from Right Accumulator		1101	XAC
		Add Index Register		1110	ADX

Table 12.1: XD-1/XD-2 instruction set in 1955 (see [IBM CCS XD][p. 65]

12.2.1 Miscellaneous class

This group contains instructions which, as the name implies, do not fit readily into other classes, including the PER instruction implemented by the selection element.[5]

HLT: This instruction halts operation of the central computer. If an input/output operation is still pending as denoted by the *IO interlock*, the instruction is delayed until completion of this operation.

---
[4][IBM AN/FSQ-7][p. 150] calls this instruction *"Branch and Index"*.
[5]See section 6.3.

ETR: The *extract* instruction is, in fact, a bitwise AND operation and implements `AC := AC & mem[address + index]`. The operand's address is contained in the right half-word of the instruction and can be indexed. Since not only the instruction word, but an operand value must be fetched from main memory, the overall execution time for this instruction is $12\mu s$.

PER: Using the *operate* instruction, input/output devices[6] can be controlled. Execution time is $12\mu s$.

CSW: The *clear and subtract word counter* instruction transfers the contents of the input/output word counter to the right accumulator in $6\mu s$.

SLR: *Shift left and round* shifts the left and right accumulator and B register simultaneously n places to the left. Each accumulator/B register pair is treated as a single 32 bit shift register. Bits shifted out to the left from bit 1 of the accumulators are discarded while the respective accumulator sign bit (which is not changed by this instruction) is used to fill the B register from the right.

Both accumulators will be rounded off to 15 significant bits by this instruction: A 1 is added to a positive value in the respective AC if the sign bit of its corresponding BR is set to 1. In case of a negative number in the AC, a 1 is subtracted if the BR sign is 1.

The A and B registers of both arithmetic units are set to a positive zero, i.e. `00...0`, after completion of this instruction which takes a variable time to complete, depending on n.

LDB: The *Load B register* instruction loads the B register with the operand fetched from the specified memory address (indexable): `BR := mem[address + index]`

Illegal instructions: All other instruction variations of the miscellaneous class have no effect other than requiring $6\mu s$ before the next instruction is executed.

12.2.2 Add class

CAD: The *clear and add* instruction effectively loads the accumulator with the operand specified, requiring $12\mu s$: `AC := mem[address + index]`

ADD: This instruction adds the left and right halves of the operand fetched from memory to the left and right accumulators. The A registers are set to +0, execution time is $12\mu s$, and an overflow may result from this instruction.

TAD: The difference between ADD and the *twin and add* instruction is that the latter adds the contents of the left half-word read from memory to the left and right accumulator. The A registers are reset to +0, and an overflow may occur.

ADB: The *add B registers to accumulators* instruction requires $12\mu s$ to complete. The A registers are reset to +0, and an overflow condition may result.

---

[6]These are in particular the line printer, card punch and reader, tape drives, area discriminators, digital displays, a situation display camera, condition lights, intercommunication indicators, the TPG, the IO interlock, and the marginal checking system.

12.2 Instruction set

CSU: *Clear and subtract* works similarly to CAD but the accumulators are loaded with the complement of the operand specified:[7] AC := !mem[address + index]. The execution time is $12\mu s$.

SUB: This instruction subtracts the left and right half-words of the operand specified from the left and right accumulators. An overflow may occur, execution time is $12\mu s$, and the A registers are set to +0.

TSU: Similarly to TAD, the *twin and subtract* instruction subtracts the left half-word of its operand from both accumulators in parallel, requiring $12\mu s$ to complete. The A registers are set to +0, and an overflow may occur.

CAM: The *clear and add magnitudes* instruction loads both accumulators with the positive left and right magnitudes of its operand. The A registers are set to +0; execution time is $12\mu s$.

DIM: *Difference magnitudes* subtracts the positive magnitudes of the left and right half-words of the operand from the accumulators, leaving the A registers set to +0, while the B registers contain the original contents of the accumulators. This instruction requires $12\mu s$ to complete.

Illegal instructions: All other add class variations are illegal instructions. Their execution adds the contents of the left half-word of the memory location specified by the operand address plus index register contents to be added to the left accumulator which may cause an overflow. The AR is set to +0.

12.2.3 Multiply class

MUL: The *multiply* instruction multiplies the contents of the left and right accumulators with the corresponding left and right half-words of the operand fetched from mem[address + index]. The magnitudes of the two operand halves are placed in the A registers, both B register bits at position 15 will contain a copy of the sign bit of their associated accumulator. Thanks to the implicit shift operation performed by the arithmetic elements, execution time for a multiplication is only $17 \pm 0.5\mu s$.

TMU: *Twin and multiply* works similar to the MUL instruction with the exception that both accumulators are multiplied with the same value, namely the left half-word of the operand fetched from memory.

DVD: The *divide* instruction divides contents of the left and right accumulators by the left and right half-words of mem[address + index]. After completion of this instruction, which requires $51\mu s$ or $52\mu s$, the magnitudes of the quotients are left in the B registers, while the remainders of the two divisions are stored in the respective accumulators. Remainders and corresponding quotients have equal signs.

---

[7] Due to using one's complement, this is the negative of the operand.

One quirk should be noted: If bit 15 of the right B register is set to 1 after the division, the complement of the right half-word of the operand remains in the A register which will contain the uncomplemented right half-word of the operand otherwise.

TDV: *Twin and divide* works quite like DVD but both accumulator contents are divided by the left half-word of the operand.

Illegal instructions: The illegal instructions of this class, each requiring $12\,\mu s$ to complete, are the most complex ones in the computer system: If the left sign bit of the operand mem[address + index] is set and if bits 10 to 15 of the right half-word of the instruction word contain 000000 or 000001, then the left accumulator and left B register will be complemented.

If the operand's left sign bit is not set, the computer will be halted if the bits 10 to 15 of the right half-word of the instruction are either 000000 or 000001. Otherwise nothing at all will happen.

12.2.4 Store class

FST: The *full store* instruction, often just called *store*, causes the contents of the accumulators to be written to the memory location addressed by address + index. Execution time is $12\,\mu s$.

LST: *Left store* stores the contents of the left accumulator in the left half-word of the memory addressed by address + index. The right-half word of this memory location will not be changed. Execution time is $18\,\mu s$ since the value of the destination address must first be read from memory to preserve the contents of its right half-word.

RST: *Right store* works the same as LST but the contents of the right accumulator will be stored in the right half-word of the destination memory location ($18\,\mu s$).

STA: From today's perspective, the *store address* instruction is quite noteworthy since it is used to modify the operand address of an instruction at runtime! It stores the contents of the right A register into the right half-word of the instruction word stored at address + index and requires $18\,\mu s$ for completion.

AOR: The *add one right* instruction increments the contents of the right half-word at memory location address + index while leaving the left half-word unchanged. After completion, which takes $18\,\mu s$, the resulting value is also stored in the right accumulator. This instruction may cause an overflow.

ECH: *Exchanges* the contents of the accumulators with the contents of memory location address + index. The original contents of this location are left in the A register. This instruction also takes $18\,\mu s$ to complete.

DEP: The *deposit* instruction stores the contents of the accumulators in a bit-wise fashion to memory location address + index. Only those bits in the destination are overwritten with bits from the accumulators where the corresponding bits of the

12.2 Instruction set

B registers are set to 1. After completion (18μs) the accumulators hold the same value as the destination location in memory: AC := mem[address + index] := (mem[address + index] & !B) | (AC & B)

Illegal instructions: All other instruction variations of this class will perform the operation mem[address + index] := +0 in 18μs. One of these instructions was later given the mnemonic STZ, as DAVID E. CASTEEL remembers.

12.2.5 Shift class

DSL: The *shift left* instruction causes a left shift of n places of the left and right accumulators and B registers (each accumulator/B register pair is treated as a single 32 bit entity). The accumulator sign bits are exempt from the shift. Bits shifted out to the left are discarded while the register pair AC-BR is filled bit-wise from the right with the value of the corresponding accumulator sign bit. Execution time depends on n, but is at least 6μs.

DSR: *Shift right* works like DSL with the difference that the contents of both AC-BR register pairs are shifted n places to the right, discarding bits shifted out to the right and filling from the left with the corresponding values of the AC sign bits.

ASL: The *accumulator shift left* instruction shift the contents of the left and right accumulators n places to the left, filling from the right with the value of the respective accumulator's sign bit, and discarding bits shifted out to the left. Execution time is variable, too, but at least 6μs.

ASR: This instruction is similar to ASL but shifts n places to the right.

LSR: The *left element shift right* affects only the AC-BR register pair of the left arithmetic element. Otherwise it works similarly to DSR.

RSR: *Right element shift right* works like LSR but affects only the right arithmetic element.

DCL: The *cycle left* instruction rotates the contents of the left and right AC-BR register pair – each treated as a single 32 bit register during this operation – n places to the left. Bits shifted out to the left are inserted again at the rightmost position.

FCL: *Cycle accumulators left* works similarly to DCL but only the two accumulators are rotated n places to the left, while the B registers are not involved.

Illegal instructions: The remaining eight instruction variations of the shift class will cause the computer to do nothing for a time depending on the value n stored in the right half-word of the instruction. Accordingly, one of these illegal instructions was later named NOP.

12.2.6 Branch class

BPX: Of the various branch instructions, the *branch on positive index* instruction is the most versatile: If the contents of the index register specified are positive or if no index register (index bits 000) or the right accumulator (index bits 011) are specified, a branch is executed.

If the branch is taken, the current contents of the program counter, already pointing to the following instruction, will be placed in the right A register and the right half-word of the *BPX* instruction will be stored into the program counter thus performing the actual branch.

If no branch is performed – due to a negative value in the index register specified or due to the index bits set to 110 or 111 – the right A register will be cleared.

The *BPX* instruction may also modify the contents of the index register specified as shown in the example program in section 12.5.4. Execution time for this instruction is $6\mu s$ in each case.

BSN: The *branch and sense* instruction is used to test for a variety of conditions like accumulator overflows, machine status etc. as specified by the *sense code* bits 10 to 15 of the left half-word of the instruction.[8] If the condition tested is true, the program counter (already incremented to point to the following instruction) will be stored in the A register while the contents of the right-half word of the *BSN* instruction will be written to the program counter. This instruction requires $12\mu s$ to complete.

BFZ: *Branch on full zero* performs a branch if both accumulators contain any combination of +0 and -0. Execution time is $12\mu s$.

BFM: *Branch on full minus* branches if both accumulators contain negative values ($6\mu s$).

BLM: The *branch on left minus* instruction performs a branch if the left accumulator contains a negative value as determined by its sign bit. Execution time is only $6\mu s$.

BRM: *Branch on right minus* works similarly to *BLM* but tests the right accumulator for a negative value. Execution time is $6\mu s$.

Illegal instructions: The illegal instructions of this class set both A registers to +0, requiring $6\mu s$.

12.2.7 Input/output class

LDC: The *load input/output address counter* instruction transfers the right half-word of its instruction word to the input/output address counter, thus specifying the first memory location from or to which a data transfer will occur. If the IO interlock is set, the instruction is delayed until the preceding input/output operation has been completed. Execution time is $6\mu s$.

---

[8] A list of all sense codes can be found in [IBM PGM][p. 401 f.].

SDR: If an input/output operation involving the magnetic drums is to be set up, the *select drum* instruction is used to specify a particular drum, field, and mode of transfer for the following IO transfer.[9] This instruction requires $12\mu s$ to complete. If the IO interlock is set, the instruction is suspended until completion of the preceding input/output operation.

SEL: Using the *select* instruction, a particular input/output is selected for the next data transfer. Possible units are the *burst* and *G/A elapsed time counters*, both part of the output system, card reader and punch, the IO register, line printer, one out of the six available tape units, the MI, and the warning light system. If the IO interlock is set, the instruction is delayed until completion of the previous input/output operation. Execution time is $12\mu s$.

RDS: The *read* instruction initiates the break-in transfer of n words from an input unit to main memory. All preceding input/output operations must be completed before this instruction will be executed. Execution time, excluding the actual transfer of data, is $6\mu s$.

WRT: The *write* instruction initiates the automatic transfer of n words as specified by its right half-word to the selected output unit. The actual transfer is controlled by the IO control element and takes place as a number of breakout cycles. Any previous input/output operation must have been completed. Execution time, excluding the actual data transfer, is $6\mu s$.

Illegal instructions: The non-assigned instructions of this class perform no operation at all, requiring $6\mu s$ to complete. If the IO interlock is set, their execution will be suspended until the outstanding input/output operation has been completed.

12.2.8 Reset class

XIN: The *reset index register* instruction sets the index register specified to the value contained in the right half-word of the instruction, requiring $6\mu s$ for execution.

XAC: *Reset index register from right accumulator* loads the index register specified with the contents of the right accumulator. Execution time is $6\mu s$.

ADX: The *add index register* instruction adds the contents of the index register specified to the right half-word of the instruction, storing the result in the right A register. An overflow is possible but can not be detected by a BSN instruction in contrast to other types of overflow. Execution time is $6\mu s$.

Illegal instructions: All remaining 13 variations of this class will just wait for $6\mu s$.

12.3 Indexed addressing

The AN/FSQ-7 architecture allows operand address modification by four index registers and the right accumulator as shown in figure 12.2. If an index register is specified

---

[9]See [IBM PGM][pp. 403 ff.] for a complete list of possible drum fields and access modes.

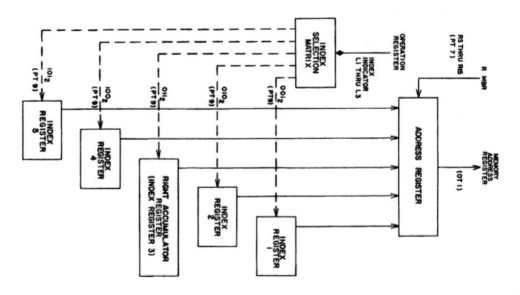

Figure 12.2: Indexed addressing (see [IBM PGM][p. 45])

with an instruction not supporting indexed addressing, it will have no effect. Specifying **000** as the index register in an indexable instruction will just use the operand address without further modification. The only exception here is the **BPX** instruction which will branch unconditionally when either **000** or **011** are specified as its index register.

12.4   Subroutines

A typical program normally requires the use of so-called *subroutines*, i. e. instruction sequences implementing some task like calculating a mathematical function or performing an intricate input/output operation. Today, such subroutines are called with special branch or jump instructions which will at least place the address to which the program flow will return upon completion of the subroutine onto a *LIFO*[10] data structure, called a *stack*. In most cases such a stack, often the same stack as that used for storing return addresses, is also used today to hold local variables of a subroutine and to pass parameters between the calling routine and the callee.

Back in the days of AN/FSQ-7 these techniques were not yet adopted, as the general concept of a subroutine had just emerged a couple of years ago. It was still distinguished between so-called *open* and *closed* subroutines.[11] Open subroutines are just sequences in a program performing a specific task but are not called from other places within the surrounding program. Accordingly open subroutines can not be used re-

---

[10]Short for *last in first out*.
[11]See [WILKES et al. 1951][pp. 22 ff.] and [IBM PGM][pp. 117 f.]. The naming convention is due to MAURICE VINCENT WILKES, 06/26/1913–11/29/2010, and his co-workers.

peatedly in a program without having multiple copies of them. In contrast to this, the closed subroutine resembles what would today just be called a subroutine: A code sequence that can be called from any place within a program which will return to the instruction following this call at the completion of the routine.

Without a stack holding at least the program address to return to, another scheme had to be used in AN/FSQ-7 and many of its contemporary and even later systems: Calling a subroutine was done using a so-called *leave provision*. This is roughly equivalent to today's *jump to subroutine* instructions and was normally implemented by a non-indexed *BPX* instruction.

This instruction places the contents of the program counter, which had already been incremented at this time to point to the instruction following the *BPX* instruction, into the right A register, before performing the actual branch to the destination address. A typical subroutine uses another *BPX* instruction at its end to return to the calling program part. Accordingly, the first instruction of such a subroutine has to store the contents of the right A register to the right half-word of its final *BPX* instruction by means of a *STA* instruction, thus making sure that this branch would return to the right location. A typical subroutine together with the call to it looks like this:[12]

```
 ─────────────────── Subroutine structure ───────────────────
1  n     BPX   m      Jump to subroutine starting at location m
2  n+1   ...          Next instruction in the calling program
3  m     STA   m+1    Store the return address in the right
4                     half-word of the instruction at memory
5                     location m+1
6  ...                Perform the task of the subroutine
7  m+1   BPX          Return to address n+1
 ─────────────────── Subroutine structure ───────────────────
```

The combination of the initial *STA* instruction and the final *BPX* operation is called the *return provision*. Since subroutines proved to be invaluable tools when it came to slicing complex tasks into smaller units of work, programs emerged which did not do much more than just calling a sequence of subroutines, anticipating the idea of threaded code. Programs of such a structure were called *master program*, *executive routine* or *sequence selection program*.[13]

12.5 Examples

Nothing is more useful when it comes to understanding a certain computer architecture and its peculiarities, than a number of programming examples. The following code fragments have been taken from [IBM PGM] and [IBM PGM 1959] and show some typical programming techniques used in the days of AN/FSQ-7.[14]

---

[12] Of course, this technique does allow recursive subroutine calls without special provisions for storing return addresses on a stack-like structure.

[13] See [IBM PGM][p. 117].

[14] Addresses and values are specified in octal notation in contrast to modern assembler programs which use hexadecimal values. Octal values run from 0 to 7, representing the bit triplets 000 to 111.

12.5.1 Polynomial evaluation

The first example program[15] is very straightforward and evaluates a polynomial of the form $y = ax^2 + bx + c$ for a given set of coefficients a, b, and c and given x by applying HORNER's[16] method. The necessary multiplication operations require some scaling to make sure that no (temporary) result will be outside the allowed range of values given the fixed point number representation scheme of AN/FSQ-7.[17]

```
                    ─── Evaluation of a polynomial ───
1   1    CAD    20      Load accumulator with a
2   2    MUL    23      Multiply this with x
3   3    DSR    n       Scale down result by 2^-n
4   4    SLR    0       Round off product to 15 bits
5   5    ADD    21      Add b to scaled value ax
6   6    MUL    23      Multiply with x again
7   7    SLR    m       Scale up by 2^m to match scale of c
8                       and round off to 15 bits
9   10   ADD    22      Add c
10  11   FST    30      Store the result y into location 30
11  12   HLT            Stop computation.
12  20    a
13  21    b
14  22    c
15  23    x
16  30   ...            Result
                    ─── Evaluation of a polynomial ───
```

First, the accumulator is loaded with the first coefficient a. Since memory location 20 contains only a 16 bit value, the right accumulator will be set to whatever is in the right half-word of this memory call (normally this would be 0). Accordingly, only the left arithmetic element is used in the remaining program although the right AE could be operated in parallel, performing a second evaluation of a polynomial. The MUL in location 2 computes ax, which has to be scaled down by suitably shifting right with DSR n. The following SLR 0 performs no actual shift operation but rounds off the result.

The next instruction, ADD 21, yields $ax + b$ in the left accumulator, which is then multiplied again by x with MUL 23, followed by a suitable scaling operation since $-1 \leq x \leq 1$ due to the fixed point one's complement number representation. Adding c finally yields $ax^2 + bx + c$, which is stored in the memory location 30 with FST 30.

12.5.2 Coordinate transformation

The next example program[18] shows the benefit of having two arithmetic elements capable of operating in parallel on different data. The task to be solved is the transforma-

---

[15] See [IBM PGM][p. 104].
[16] This method is named after WILLIAM GEORGE HORNER, 1786–09/22/1837.
[17] See [IBM PGM][pp. 427 ff.] for detailed information about scaling.
[18] See [IBM PGM][p. 103].

12.5 Examples

tion of polar coordinates to Cartesian coordinates, an operation required abundantly in the software running the DCs and CCs.

A radar station located at a known location described by constant Cartesian coordinates x_r and y_r, with respect to some common point of reference,[19] sends a pair of polar coordinates R and φ, representing range and azimuth angle, to the DC. These coordinates denote the position of a detected target with the radar station being in the origin of the polar coordinate system. The task is now to determine the target position in Cartesian coordinates x_t and y_t taking the common point of reference into account:

$$x_t = x_r + r\cos(\varphi) \text{ and}$$
$$y_t = y_r + r\sin(\varphi).$$

Such a pair of coordinates could then be used to display the target on a situation console or to guide interceptors to the target.

```
―――――――――――――――――― Coordinate conversion ――――――――――――
 1   1              CAD       30        Load φ in right AC and R in left AC
 2   2          3   CAD       100       Load cos(φ), sin(φ) into accumulators
 3   3              TMU       30        Multiply both with R
 4   4              SLR       0         Round products to 15 significant bits
 5   5              ADD       20        x_r + R cos(φ) and y_r + R sin(φ)
 6   6              FST       30        Replace R, φ in memory with these values
 7   7              HLT                 End of program
 8   20             x_r       y_r       Input data
 9   30             R         φ         Input data, overwritten by result
10   100            cos(0)    sin(0)    Lookup table
11   ...
12   100 + N        cos(N)    sin(N)
―――――――――――――――――― Coordinate conversion ――――――――――――
```

The first instruction loads the value pair R and φ as received from the radar station and stored in memory location 30 into the left and right accumulators. The following CAD-instruction, indexed with the right accumulator according to figure 12.2, performs a table lookup based on the value φ to load the values $\sin(\varphi)$ and $\cos(\varphi)$ required for the actual conversion from polar to rectangular coordinates into the two accumulators. These are then both multiplied by R with TMU 30 and SLR 0, so that the accumulators now hold the rectangular target coordinates $R\sin(\varphi)$ and $R\cos(\varphi)$ with reference to the radar station. Adding x_r and y_r with ADD 20 yields the desired coordinates x_t and y_t, which are then stored into memory location 30 with FST 30.

---

[19] Obviously, knowing the exact location of every radar station connected to the DCs was of prime importance and not a simple task at that: *"[W]e had to align the radars, and we therefore had to know where they were. We sent surveying teams out to find their locations. They surveyed all the radars, and we cranked up the system and discovered the radars weren't registering. We eventually discovered that sometimes there were mistakes of miles in the surveyed locations. The location given for one radar turned out to be in the middle of Long Island sound."* See [TROPP et al. 1983][p. 397].

12.5.3 Finding the largest number

The two example programs above did not need any branches at all. The next program[20] determines the largest of four given values $n_1, \ldots, n_4$ requiring a lot of conditional branches while using only the left accumulator again:

---------- Find largest ----------

#	Addr	Op	Arg	Comment
1	1	CAD	30	Load n_1 into accumulator
2	2	DIM	31	Subtract n_2
3	3	BLM	15	If result is negative ($n_2 \geq n_1$), branch to 15
4	4	DCL	20	$n_1 > n_2$: Restore n_1 from B register
5	5	DIM	32	Subtract n_3
6	6	BLM	17	Branch to 17 if result is negative
7	7	DCL	20	Restore last value from B register
8	10	DIM	33	Subtract n_4
9	11	BLM	21	Branch to 21 if result is negative
10	12	DCL	20	Restore last value from B register
11	13	LST	40	Store content of left accumulator to address 40
12	14	HLT		Halt
13	15	CAD	31	Load n_2 into accumulator
14	16	0 BPX	5	Branch to address 5
15	17	CAD	32	Load n_3 into accumulator
16	20	0 BPX	10	Branch to address 10
17	21	CAD	33	Load n_r into accumulator
18	22	0 BPX	13	Branch to address 13
19	30	n_1		
20	31	n_2		
21	32	n_3		
22	33	n_4		
23	40	...		Result

---------- Find largest ----------

The basic idea is straightforward: First the value n_1 is loaded into the left accumulator.[21] Then the magnitude of n_2 is subtracted from the accumulator by DIM 31. If the result is negative, i.e. $n_2 \geq n_1$, the accumulator is overwritten with n_2 by branching to the instruction at address 15. If the result is not negative, the branch is not taken and the original value n_1 is restored in the accumulator. Instead of loading the accumulator again from memory, the register pair AC-BR is cycled 16 positions to the left[22], effectively restoring the original contents of the accumulator since the DIM instruction stores the original contents of the accumulator in the associated B register before performing the subtraction.

The accumulator, now holding either n_1 or n_2, depending on what value was larger, is then compared against n_3 by DIM 32, followed by a conditional branch BLM 17 to address 17 and the same sequence of operations as described above follows. A fourth such sequence eventually leaves the biggest of the four values in the left accumulator.

---

[20] See [IBM PGM][p. 105].
[21] The right arithmetic element is running in parallel working on zero data.
[22] 20 in octal notation as used in the program listing.

12.5.4 Adding ten numbers

The next problem involves address modification. The problem to be solved is adding a number of value pairs stored in consecutive memory locations using both arithmetic elements in parallel.[23] The first implementation uses address modification, essentially modifying the address part of an instruction during run time:

```
                      ── Adding ten numbers -- iterative version ──
1     1    CAD    30    Load first number into accumulator
2     2    BPX     4    Unconditional branch to first add
3     3    CAD   150    Load last partial sum into accumulator
4     4    ADD    31    Add next number to partial sum
5     5    FST   150    Store partial sum into address 150
6     6    AOR     4    Modify address part of the instruction at
7                       address 4 by adding one. Modified address
8                       is left in the right accumulator
9     7    SUB   147    Compare this address with the last data address
10   10    BRM     3    Repeat adding sequence unless completed
11   11    HLT          Halt
12   30    $n_1$        $m_1$    Numbers to be accumulated
13   ...
14   41    $n_{10}$     $m_{10}$
15  147          41     Address of last number to be added
16  150         ...     Temporary and final result storage
                      ── Adding ten numbers -- iterative version ──
```

CAD 30 loads the left and right accumulators with the first value pair n_1, m_1. The following instruction BPX 4 is an unconditional branch to address 4, since no index register is specified, initially skipping the instruction at address 3. The instruction at address 4, ADD 31, then adds the contents of memory location 31 to the accumulators. The following FST 150 stores this (partial) result at address 150 for later use.

The next instruction is unusual from today's perspective: AOR 4 increments the address part of the instruction stored at address 4, changing it to ADD 32 etc. Since AOR leaves the incremented address in the right accumulator, this can be used to determine if all values have been added already. Subtracting the last address of data stored at address 41 from the right accumulator by SUB 147 yields a negative result if the last value pair has already been added. The BRM 3 instruction accordingly branches back to address 3 only when there are values left to be added. The CAD 150 instruction loads the last partial sum into the two accumulators.

Obviously such a self-modifying program can not be called twice without restoring all altered address parts of its instructions – a serious drawback. In addition to this, using AOR in a place where an index register might have been used is neither overly elegant nor efficient. Accordingly, the AOR was more often used to increment counters stored in right half-words in memory instead of performing address modification.[24] The following variant solves the same problem using indexed addressing:

---
[23]See [IBM PGM][pp. 109 f.].
[24]See [IBM PGM 1959][p. 53].

―――― Adding ten numbers -- indexed version ――――

1	1	1 XIN	10	Set index register 1 to decimal 8
2	2	CAD	30	Load first number into accumulator
3	3	1 ADD	31	Add next number (address 31 + content
4				of index register 1) to accumulator
5	4	1 BPX(01)	3	Branch to address 3 if index register 1
6				is positive, and subtract index interval
7				from index register 1
8	5	FST	150	Store result into address 150
9	6	HLT		Halt
10	30	n_1	m_1	Numbers to be accumulated
11	...			
12	41	n_{10}	m_{10}	
13	150			Temporary and final result storage

―――― Adding ten numbers -- indexed version ――――

The first instruction, 1 XIN 10, sets index register 1 to the octal value 10 (eight in decimal). The following CAD 30 loads both accumulators with the first value pair stored at address 30. The loop, consisting of the instructions at addresses 3 and 4, first adds the next value pair as addressed by 31 + ir[1][25] by 1 ADD 31 before executing the instruction 1 BPX(01) 3 which requires a bit of attention as it performs two operations at once: First it tests if the contents of index register 1 are positive before subtracting 1 from the index register. If the test for positiveness came out true, a branch back to address 3 is performed.

The first run of this loop adds the value pair stored at address 41[26], the next run through the loops adds the contents of address 40 etc. Since the 1 BPX(01) 3 instruction tests the contents of the index register *before* decrementing it, the loop is actually executed nine times although index register 1 had been loaded with octal 10 initially!

12.5.5 Delaying

In some cases a programmed delay is quite useful. The following program shows a typical routine causing a $120\mu s$ delay:[27]

―――― Delay of $120\mu s$ ――――

1	0	5 XIN	22	Load index register 5 with 22
2	1	5 BPX(01)	2	Branches to itself until index register
3				5 becomes negative, decrement index
4				register 5
5	2	HLT		

―――― Delay of $120\mu s$ ――――

―――――――――――――

[25] ir[1] denotes index register 1.
[26] Octal 31 + 8.
[27] See [IBM PGM 1959][pp. 57 f.].

5 XIN 22, which loads index register 5 with the octal value 22, takes 6µs to complete, the tight loop 5 BPX(01) 2 is executed 19 times[28], requiring another 6µs per iteration for a total of 120µs.

12.5.6 Printing

The following, last example program sends data to the line printer:[29]

———————— Printing ————————
1	SEL (03)		Select line printer for output
2	LDC	i	Load IO address counter with address of
			first value to be sent to the printer
3	BSN (11)	j	Check if printer is not ready
4	WRT	k	Write k words to the printer
	...		
j	...		Error handling routine
	...		
i	...		First address of data to be printed
———————— Printing ————————

The first instruction selects the line printer as output device. If any previous input/output instruction is still pending as denoted by the IO interlock, the SEL (03) instruction will be suspended until the interlock is cleared. The next instruction, LDC i loads the IO address counter with the address of the first data word to be sent to the printer. Using BSN (11) j, the printer is checked for any error conditions preventing a successful operation.[30] The actual data transmission of k data words, which is performed using breakout cycles as described in section 6.3, is initiated with the WRT k instruction.

12.5.7 Trick programs

Every computer, especially a large and complex one as the AN/FSQ-7, invites clever programmers to try things not suggested by the programming manuals. One interesting program that unfortunately has been lost, is remembered by DAVID E. CASTEEL:[31]

> "I once took advantage of the format of the binary load card[32] and idiosyncrasies of the instruction language to write a program that could be punched on a single IBM card that would execute and perform a different simple task when read into the computer in each of its 4 possible orientations. (It caused a different one of the 4 condition lights[33] to blink on and off depending on which way it was read in.)

---

[28] As before, the loop is executed once more than one might expect due to the fact that BPX tests the contents of the index register before decrementing it.
[29] See [IBM PGM][p. 87].
[30] Typical error conditions for a line printer are missing paper, being offline due to the start-button not having been pressed, or an incorrectly wired control panel.
[31] Its recreation may be left as an exercise to the reader.
[32] A punch card containing binary data as described in section 8.5.
[33] These software controlled condition lights were located on the maintenance console.

> *I do recall that it depended a great deal on the fact that so many operation codes were essentially [a] NOP[34] and so those read backwards were simply ignored. For all practical purposes the card just contained 4 separate 6-step programs, only one of which would actually do anything depending on what orientation it was read in. The layout of the card involved 24 32-bit words in 2 columns [...] with a column of 16 bits left over that was used normally for identification/sequence numbering. When the card was inverted, the null space became the address portion of instructions and all the other half-words had their meanings swapped – addresses became instructions and instructions became addresses; one set of addresses became a new null space, too. The combination of so many NOP instruction codes and the very redundant address scheme for little memory[35] made it possible to find codes that could work for either purpose."*[36]

As short as all of the preceding programming examples are, they give a good impression of the programming techniques used in the 1950s and early 1960s and the intricacies of the AN/FSQ-7 computer architecture with its one's complement fixed point number representation, dual arithmetic elements and its IO element. The following chapter describes the operational software ran at DCs and CCs.

---

[34]Short for *no operation*.
[35]This denotes the 4 k core memory subsystem.
[36]DAVID E. CASTEEL, Captain, USAF (ret), personal communication.

13 Software

Developing the necessary software for the DCs and CCs proved to be a formidable task of hitherto unknown complexity; a task that was quite underestimated at the very beginning of the development of SAGE. This was further complicated by the fact that there was no "software industry" at all:

> "It was [...] difficult with the software. We had by then written the Cape Cod programs and had some feeling for the difficulty. We tried to get IBM interested in it and their answer was, 'No, we sell equipment.' So we tried AT&T who declined. Finally, System Development Corporation, spun off from the Rand Corporation, was created for this purpose. [...] The software turned out to take thousands of people. JAY set up a recruiting operation, and we hired hundreds of people off the street, unemployed mathematics teachers and so on. The Lincoln group hired hundreds of people for SDC."[1]

According to numerous rumors, about 20% of all programmers available worldwide worked on SAGE during its heyday, developing the programs running the DCs and CCs, which were called DCA[2] and CCA[3] respectively. The complexity of these developments is shown by the fact that bringing the first AN/FSQ-7 installation at McGuire AFB to operational status from a software perspective took two years.[4] The sheer size and complexity of a typical DCA is reflected by the following quotation:[5]

> "An active DCA program which SDC sends to an operational site averages 4 reels of magnetic tape and approximately 100,000 cards. When all associated documentation is provided, it is estimated to amount to 446,000 pages weighing about 3200 pounds."[6]

13.1 Software development process

One of the lessons learned early in the process of developing the software that would eventually form the heart of the widespread SAGE system with its numerous computer

---

[1] See [EVERETT 1988][p.14].

[2] Short for *direction center active*.

[3] Short for *combat center active*. The CCA ran on AN/FSQ-8 installations with their reduced input/output equipment.

[4] The equipment had been installed in 1956 at McGuire AFB, but operational status was achieved only in 1958.

[5] The only other early system comparable in complexity to the DCA and CCA may have been the software for the Apollo guidance computer, see [O'BRIEN 2010].

[6] See [ADC SDC][p. 12].

PROGRAM	LENGTH
Compiler	10,500
Read-in	1,300
Library Merge-Output	4,700
Checker	7,500
Master Tape Load	2,000
In-out Editors	2,400
Communication Pool	4,100
Utility Control	3,000
Numeric Subroutines	1,000
Miscellaneous	4,000
	40,500

Table 13.1: Breakdown of the Lincoln Utility System (see [BENINGTON 1983][p. 359].)

centers, its myriad of connected radar stations, missile and interceptor bases, was that software is hard. As successful as the early experiments with the Cape Cod system were, they lulled nearly all persons involved into a false sense of security. However, the bitter awakening came when it turned out that these rather simple programs – as complex as they were, they were at least small enough to be fully understood by a single person – would not scale easily from a single or two-computer installation to a nationwide large scale system.

This might have been the first time in human history that programs had to be developed which were far too complex for a single person to grasp and understand every detail of the code. A further complication was the result of the real-time nature of the operation of SAGE – events came in from adjacent radar stations and other installations with no foreseeable pattern etc. All of this had to be solved with a machine of – even back then – very limited capacity and capabilities. With only two times 4 k of main memory, later expanded to 4 k plus 64 k of memory, and a speed of about 75,000 instructions per second, the hardware proved to be a challenge in itself. Squeezing the highly complex software in such a constrained environment alone would have been a formidable task.

To accomplish this Herculean task, a variety of tools had to be created, including libraries of subroutines for handling numerical computations, input/output etc., compilers, linkers and loaders, etc. Table 13.1 gives an impression of the tools developed at Lincoln laboratory during the development of the operational software for SAGE. According to [BENINGTON 1983][p. 360], another 10,000 instructions for additional special programs, 20,000 instructions for test instrumentation and 30,000 instructions for operational instrumentation – all of this on top of the 40,500 instructions mentioned in table 13.1 – were eventually required.

The development of this vast amount of software required – for the first time in history – a software development process, which was first described in 1956 in a presentation by HERBERT D. BENINGTON at a symposium on advanced programming methods for digital computers.[7] This presentation was aptly titled *"Production of Large Computer*

---

[7] See [BENINGTON 1983].

Programs" and is an early foray into what would eventually evolve as the discipline of *software engineering*. The process of *program construction* described in BENINGTON's paper is based on nine sequential steps, beginning with an *operational plan*, followed by *machine* and *operational specifications*.[8] These were followed by *coding specifications*, the actual process of *coding, parameter* and *assembly testing*, a *shakedown*, and finally a *system evaluation*.[9]

Eventually, the software development process was carried out under the auspices of the SDC.[10] All development actions were bundled into the following three groups:[11]

System design: Consists of the *"operational analysis"* during which the system requirements are determined. These requirements are then detailed and studied to yield *"conceptual solutions"* for a given problem. These are then verified in a *"concept verification"* step. These steps make extensive use of mathematical and simulation modeling as well as mathematical analysis.

Operational design: Provides detailed descriptions of operational functions for the SAGE DCs and CCs. This is an iterative process in which ideas are generated, documented and discussed over and over again before they are finalized and written down in the operational design specifications. All in all there are 25 functional areas, all heavily interdependent, like weapons assignment and direction, radar data input, tracking, etc. up to the design of the consoles, which have to be taken into account.

The operational design process consists of six subprocesses: The *"requirements phase"* during which the effects of the change in requirements on the overall system are studied, the *"preliminary design phase"* and the *"pre-concurrence phase"* where initial specifications are written down and reviewed by all stakeholders, and the *"concurrence phase"* which is basically a formal meeting involving all affected agencies, civilian as well as military. These agencies then have to concur on the proposed specifications. This is followed by the *"post design phase"* which is merely an active evaluation of the proposed design. The goal of this phase is to identify improvements caused by this recent program change. Finally, a *"documentation phase"* takes place. A documentation set consists of the *"Operational Specifications"*, the *"model*[12] *change training guide"* (contains the differences between the old and new program models with regard to their training implications), *"specific design studies"* where necessary, the *"SAGE Positional Handbooks"* describing all operational procedures, responsibilities etc., and, finally, the *"SAGE System Description"* describing the DC and CC programs and their respective operation.

---

[8] Given today's nearly monocultural hardware environment, a machine specification feels a bit awkward from a programmer's perspective.

[9] This rather sequential approach is obviously a forerunner of the so-called *waterfall model*.

[10] [JOHNSON 2002][pp. 156 ff.] contains a thorough description of the technical as well as the managerial challenges of programming a highly complex, interconnected and mission-critical system such as SAGE. [JACOBS 1986] gives a highly interesting personal account of the software development process for SAGE.

[11] See [ADC SDC][pp. 18 ff.].

[12] The term *model* denotes what would be called a *version* or *revision* in modern parlance.

Program design: This final step consists of five phases: *"Program costing"* (determination of new features and their implications regarding storage capacity, processing time etc.) – as a result, schedules are determined. The next phase is the *"program design phase"* which covers both coarse and fine design; features are assigned to coordinators taking over responsibility for these, schedules are established for coding, testing etc. The next phase is the *"coding phase"* where the actual task of programming takes place (this is an iterative phase). The following *"Test phase"* consists of *"parameter testing"* (testing of individual subprograms). Finally, groups of subprograms are tested during the *"assembly testing"*. The last phase is, again, a *"documentation phase"*.

The actual *production* of a program took place in Santa Monica at SDC and included extensive tests using simulated input data to find coding errors. Only later tests involved live air situations which were extremely costly and complex to arrange and execute. Following this, live environment testing took place at DC 08, Richards Gebaur AFB in Kansas City. Another task performed during this stage was that of *adaptation*. During this process, the model to be delivered to a certain DC or CC was adapted to its particular environment. These changes concerned the positions of the radar stations attached to a particular center, the assignment of LRI, GFI, XTL and output lines to telephone channels, changes in the controlled air defense equipment etc.

Following these adaption changes, the model was shipped to its destination centers and installed. Since the configuration parameters derived previously could be erroneous, additional *field installation testing* was required before a new software version would be brought into production.

Any errors found during operation of a certain model were reported back and found their way into the so-called *model maintenance* process which would eventually yield *SAGE Programming Changes (SPC)*s.[13] So-called *emergency changes* were performed locally at the various DC and CC sites by their respective on-site SDC field programming teams.

13.2   Operational software

The expenses for developing the complex DCA and CCA were comparable with the cost for the hardware they ran on:

> *"Let us assume an overhead factor of 100 percent (for supporting programs, management, etc.), a cost of $15,000 per engineering man-year (including overhead), and a cost of $500 per hour of computer (this is probably low since a control computer contains considerable terminal equipment). Assuming these factors, the cost of producing a 100,000-instruction program comes to about $5,500,000 or $55 per machine instruction. In other words, the time and cost required to prepare a system program are comparable with the time and cost of building the computer itself."*[14]

---
[13] See [ADC SDC][pp. 33 ff.].
[14] See [BENINGTON 1983][p. 357].

13.2 Operational software

Unfortunately, the actual size of a complete DCA is hard to determine as nearly everything seems to have been lost and no distribution tape is known to exist anymore. Early sources such as [MIT 1956][p. 26] quote a program size of 56,779 words plus 29,788 machine words for tables storing data, while later recollections such as [EVERETT et al. 1983][p. 335] contain size estimations between 75,000 and 100,000 instructions. This matches well with personal recollections such as those by DAVID E. CASTEEL who quotes a size of roughly 93,000 instructions for the DCA. These numbers seem realistic, given the hardware constraints and the available documentation etc., while others[15] seem to be way over the top with estimates of one million words for the operational software.

Today, such a real-time system would most probably rely heavily on an intricate interrupt processing system, with a real-time executive at its heart, prioritizing the processing of incoming events etc. The approach chosen for the DCAs and CCAs was as different from that as possible. Its core was called *Program Executive Control (PEC)* and resembled what today might be called a *supervisor*. The basic operations of gathering data from radar stations etc., performing the necessary computations for determining tracks and the like, the control of the various SDs and DDs, were performed in a loop over and over again with no interrupt processing at all. Most input/output was done by polling drum status fields and reading or writing data only when present or writable. As ROBERT R. EVERETT remembers,

> "[e]very $2\frac{1}{2}$ seconds, the computer generates about 200 different types of displays, requiring up to 20,000 characters, 18,000 points, and 5,000 lines. Some of these are always present on an operator's situation display. Others he may select. Some he may request the computer to prepare especially for his viewing. Finally, the computer can force very high-priority displays for his attention."[16]

The DCA was organized into so-called *frames*, each of which was further divided into three five-second *subframes*. One such subframe generated two display cycles, so the typical display update rate was 2.5 seconds. If these update intervals became too long due to excessive amounts of data to be processed, input data would be manually restricted, so that only information absolutely necessary to perform the air defense task was allowed to enter the input drums, thus alleviating the load on the machine and shortening the frame time. This frame based processing can be clearly seen in the few and rare video recordings of typical situation displays: Every now and then the large display screen flashes with a blue streak of light when the tracks, targets and additional information are refreshed during one frame. Afterwards the screen shows only the orange afterglow of its long persistence phosphor which grews dimmer and dimmer with every second.[17]

---

[15] Like [BELL 1983][p. 13].

[16] See [EVERETT et al. 1983][p. 3336].

[17] Certainly not an environment that would be called "user friendly" today. Adding to this the continuous stress caused by the burden of performing a task substantial for the nation's security, the long hours, the dim lighting, one can only imagine the physical and psychological toll this operating environment took on the myriads of operators sitting in front of those display consoles.

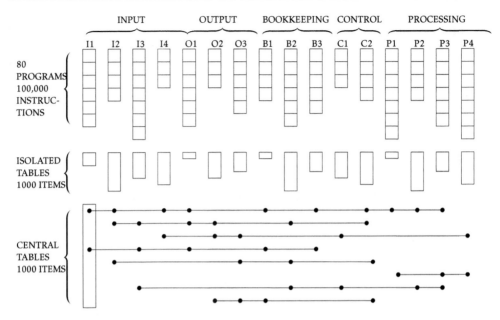

Figure 13.1: Static program organization (cf. [EVERETT et al. 1983][p. 335])

Figure 13.1 shows the static program organization of a DCA. It consists of about 80 subprograms, each between 25 to 4,000 instructions in size,[18] arranged in five groups dealing with input, output, bookkeeping, control, and general processing tasks. Subgroups of these programs share data through so-called *isolated tables*, while a set of *central tables* is used for communication between such groups of subprograms. The overall control of these subprograms is performed by a sequence control program similar to the general idea of a master program as described in section 12.4.

This architecture with its central control program, calling a variety of subroutines in a suitable sequence over and over again in conjunction with the abdication of interrupt processing yielded a highly stable software system:

> "[T]he key here is the aging and the fact that the program is NOT interrupt driven. The program simply cycles through the job queue every few seconds in a round robin fashion. This is an excellent example of superb software engineering with an incredibly simple overall structure since it is non-parallel, all the bugs that an interrupt driven system would have had are avoided. Users identify overload by the lengthened cycle time."[19]

---

[18]See [EVERETT et al. 1983][p. 338].
[19]See [BELL 1983][p. 14].

14 Failure or Success?

As impressive as SAGE with its marvelous duplex AN/FSQ-7 and AN/FSQ-8 computers was, there was always harsh criticism from various sides: Some claimed that SAGE was a waste of taxpayer's money with all of its *"gold plated"* equipment, while others called it a complete failure, incapable of performing the air defense task. Still others marvel at the technologies developed for SAGE and think of it as a good investment in the future of a whole industry. The following sections will shed some light on both sides of the coin.

14.1 A failure?

In some respect, SAGE was outdated even after its commissioning in 1958 as the main threat of air defense had already shifted from long range bombers equipped with thermonuclear weapons toward *Intercontinental Ballistic Missile (ICBM)* installations. The first of these new weapons, the A9/A10 rocket, was designed in Germany beginning in 1940, although no actual working system was built and tested. Nevertheless, the idea of a long range rocket based weapon system was prevalent since the 1940s and in 1953 development of an ICBM capable of delivering a thermonuclear warhead to a long distance target began under SERGEI PAVLOVICH KOROLEV.[1] This led to the *R-7* rocket which performed its first successful test on August 21st 1957 and reached a distance of about about 8,000 km, way enough to reach targets outside the USSR. Interestingly, General HENRY HARLEY ARNOLD[2] somewhat prophetically wrote in 1943:

> *"Someday, not too far distant, there can come streaking out of somewhere – we won't be able to hear it, it will come so fast – some kind of gadget with an explosive so powerful that one projectile will be able to wipe out completely this city of Washington."*[3]

Even the following 1947 statement of Brig. General THOMAS SARSFIELD POWER,[4] which turned out to be correct, did not neglect the looming threat of ICBMs:

> *"For the next 10 years, long-range air bombardment will be effected by means of subsonic bombers only."*[5]

---
[1] Сергей Павлович Королёв, 01/12/1907–01/14/1966
[2] 06/25/1886–01/15/1950
[3] See [CORRELL 2005].
[4] 06/18/1905–12/06/1970
[5] See [CORRELL 2005].

So it was foreseeable even in the late 1940s that the long range bombers, which were the basis of definition of the air defense problem which in turn led to the development of SAGE, were soon to be history, replaced by ICBMs against which next to no countermeasure was possible. Accordingly, the often heard accusation that SAGE was outdated from its very beginning is not without a grain of truth.[6]

Even a conventional strike of the USSR, throwing a plethora of long-range bombers against targets in the United States, would have been outside the detection and processing capabilities of SAGE. Although there were numerous test raids flown during operational weapons tests and evaluations, the realism of these scenarios may be questionable. The reported *kill percentages*, always near 100%, were obtained under ideal conditions with no electronic counter measures by an enemy such as radar *jamming* and the like.[7]

Another weak point of SAGE was the decision to build the DCs and CCs above ground.[8] This was due to a half-hearted attempt at cutting costs. Of course, structures like the SAGE blockhouses were easy targets and might also have attracted an explicit attack:[9]

> *"[...] SAGE blockhouses were conspicuous structures that dominated the landscape wherever they were located. A mere handful of enemy missiles, therefore, could severely cripple US defenses against the manned bomber."*[10]

PAUL N. EDWARDS remarks with a rather grim sense of humor that

> *"[the] decision [of building the SAGE installations above ground] had only one possible strategic rationale: [Strategic Air Command (SAC)] intended never to need SAGE warning and interception; it would strike the Russians first. After SAC's hammer blow, continental air defenses would be faced only with cleaning up a weak and probably disorganized counterstrike."*[11]

On the other hand, SAGE might not have been useless at all from a military perspective – there never was a preemptive strike of the USSR against the United States, the Cold War never became hot. So SAGE might have at least been a successful deterrence.

---

[6] Accordingly, SAGE was once called *"a marvelous technological solution to the wrong problem"*. See [EARNEST et al. 1998]. The main reason why SAGE was held into operation until 1983 may have been the fact that the USSR did not abandon their strategic bomber fleet with the advent of the ICBM.

[7] The term *jamming* describes the technique of saturating a radar receiver by intentionally emitting fake target echoes etc. It was planned to counteract jamming at the earliest stage possible, i. e. at the radar stations by prefiltering, changing frequencies (*Frequency Diversity (FD)*) etc. It was the job of the *Electronic Counter Counter Measures Technician (ECCMT)*s to counter any jamming attempt of those radar systems. SEVER M. ORNSTEIN writes *"I believe SAGE would have failed utterly in the presence of active jamming by an enemy, but fortunately it was never put to the test."* (see [ORNSTEIN 2002][p. 26]).

[8] Only the Canadian DC 31 was an underground installation.

[9] Accordingly, the DCs and CCs were often called *bonus targets*.

[10] See [WAINSTEIN et al. 1975][p. 215].

[11] See [EDWARDS 1997][p. 110].

Nevertheless, a 1969 incident clearly demonstrated the limitations of SAGE as a whole and should become the beginning of the end: On October 5th, 1969, a defecting lieutenant from Cuba landed with his MIG fighter on Homestead AFB. As embarrassing as this alone would have been, it was even worse since only a short time ago the *Air Force One* had landed on Homestead AFB with the president of the United States aboard.[12] This debacle turned into a severe blow against SAGE as congress decided to cut back financial funding substantially in the aftermath.

14.2 Success!

Despite the undisputable immense costs of SAGE, common estimates range from $4–$12 billion, with $8 billion being a probable value,[13] SAGE turned out to be a fundamental part of our technological heritage. Nearly all following developments in computer technology were pioneered in AN/FSQ-7. SAGE, in fact, created what became a computer *industry*. The countless spin-offs from this giant undertaking are so invaluable that SAGE can hardly be seen as a waste of time and money from today's perspective.

14.2.1 Hardware

From a computer architectural point of view, AN/FSQ-7 was the first large scale production machine clearly demonstrating the power of indexed addressing with its BPX instruction.[14] In addition to this, AN/FSQ-7 pioneered the idea of a dedicated input/output control unit capable of transferring blocks of data without intervention of the central processor, predating today's ubiquitous *direct memory access*[15] controllers.

By far the most important asset from Whirlwind and AN/FSQ-7 was the development of reliable large scale core memory systems. For several decades to follow this solved one of the most important problems of computer technology. Prior to these developments, memory systems were large, clumsy, and error-prone, often determining the ridiculously short up-times of early stored program digital computers. The very idea of three-dimensional stacks of magnetic cores to store binary data boosted a whole industry and also paved the way for complex and memory hungry software systems, including compilers and interpreters.

The reliable transmission of digital data over cheap voice-grade telephone lines was also pioneered by SAGE. The transmission devices, forerunners of today's *modems*,[16] became a central contribution to the Bell System *A-1 data system*[17] and clearly demonstrated the practicability of nation- and even worldwide computer networks:

---

[12]See [BRIGGS 1988][pp. 337 ff.].
[13]See [PRESS 1996][p. 12].
[14]The idea of an index register itself was not new, it had already been implemented 1948 in the *Manchester Mark I*, designed by FREDERIC CALLAND WILLIAMS and TOM KILBURN (see [ASTRAHAN et al. 1983][p. 348]).
[15]DMA for short.
[16]Short for *modulator/demodulator*.
[17]See [TROPP et al. 1983][p. 396].

> "The government also bootstrapped computer communication with Whirlwind, an *Office of Naval Research* computer that processed telemetry data in real time at MIT. Whirlwind demonstrated the feasibility of real-time data communication over analog telephone lines [...] After the Whirlwind R&D, the government turned to procurement with the Semi-Automatic Ground Environment (SAGE) system [...] SAGE was the first computer network, growing finally to link Q-7s and Q-8s in 26 centers.[18]

The AN/FSQ-7/8 systems also were the first large scale systems relying on circuit standards being employed throughout the system, ranging from basic circuits and their packaging in easily exchangeable modules, to standard mounting frames, intercabinet cabling etc. Another feature pioneered by these machines was parity checking throughout the machine, involving every internal data transfer. It should take years before commercial machines featured so-called *fully checked designs*.[19]

AN/FSQ-7 also pioneered the idea of marginal checking and, much more influential, that of duplexing computers to increase availability, becoming the forerunner of today's cluster solutions. MORTON MICHAEL ASTRAHAN[20] and JOHN FRANCIS JACOBS[21] aptly note that

> "[t]he SAGE system provides a demonstration of the kind of innovation that can be achieved when cost is secondary to performance. This kind of environment is difficult to create in a commercially oriented company, but SAGE provided the environment. Ambitious performance goals were met by the operational systems. Furthermore, as hardware costs dropped, most of the SAGE innovations became cost effective for the commercial market."[22]

14.2.2 Graphics

Figure 14.1: Pin-up displayed on a SD tube at the Fort Lee DC (©Airman First Class LAWRENCE A. TIPTON, 1959)

It should take years for other machines to reach the graphical capabilities of AN/FSQ-7 and its associated situation displays. These marvels of technology predated all of the later vector graphics displays and pioneered *graphical user interfaces, GUIs* for short, with the development of the light gun.

Of course, the availability of such a computer system resulted in not too serious programs, too. The now famous pin-up shown in figure 14.1 is "[the] world's earliest known figurative computer art, and quite possibly the first image of a human being on a computer screen." According to

[18] See [PRESS 1996][p. 12].
[19] See [TROPP et al. 1983][p. 392].
[20] 12/05/1924–06/02/1988
[21] 02/22/1923–09/12/1998
[22] See [ASTRAHAN et al. 1983][p. 347].

LAWRENCE TIPTON, the program generating the pin-up image consisted of about 100 punch cards.[23] Research done by BENJ EDWARDS shows that the picture displayed closely resembles a pin-up drawn by artist GEORGE BROWN PETTY IV[24] for a December 1956 calender page.

14.2.3 Software

Being the first large scale software development project, SAGE coined and formed a complete sector of technology. It not only demonstrated the overwhelming and always underestimated complexity of software development as such but also offered the first formalized techniques of dealing with this new degree of complexity. It gave a first foretaste of the so-called *software crisis* and would become catalyst of a new subfield in computer science eventually called *software engineering*.[25]

Apart from these indirect contributions to software technology, SAGE pioneered many ideas and techniques which we today take as granted, such as *time-sharing, real-time processing,* modular (top-down) system organization, table-driven software simplifying later modification and extension, centralized data structures,[26] etc. Especially, these central data structures turned out to be quite inspiring:

> *"By the late 1960s, however, 'data base' was a common expression used in corporate computing circles, largely replacing the hubs, buckets, and pools in which data had previously been rhetorically housed. This term was imported from the world of military command and control systems. It originated in or before 1960, probably as part of the famous SAGE anti-aircraft command and control network. [...] SAGE had to present an up-to-date and consistent representation of the various bombers, fighters and bases to all its users. The System Development Corporation, a RAND Corporation group spin-off to develop the software for SAGE, had adopted the term 'data base' to describe the shared collection of data on which all these views were based. SDC actively promoted the data base concept for military and business use. Its interest in general purpose data base systems was part of its attempt to find new markets for its unique expertise in the creation of large, interactive systems. [...SDC] identified 'computer-centered data base systems' as a key application of time-shared systems – hosting (in collaboration with military agencies) two symposia on the topic in 1964 and 1965. The SDC Data Base Symposia were crucial in spreading the data base concept beyond the world of real-time military contractors."*[27]

---

[23]Since one binary card could hold 24 machines words of 32 bits, the overall program including data occupied only about 2,400 words of storage. Some sources, such as [EDWARDS 2013], claim that this program was used as an diagnostic aid, which might be a bit of an exaggeration but it is a nice tale nevertheless. There are also rumors of yet another program which allegedly displayed a hula girl. By means of the SD switches the hula girl even danced, and pointing at her skirt with the light gun and pulling the trigger would cause the skirt to drop and the screen to go blank afterwards.

[24]04/27/1894–07/21/1975

[25]See [BENINGTON 1983].

[26]Most of the data areas used by the various subprograms of the DCA and CCA were grouped into a so-called *COMPOOL*.

[27]See [HAIGH 2006][p. 35].

14.2.4 ATC and SABRE

Another direct spin-off of SAGE is the application of networked computer systems to *air traffic control*.[28] The MITRE Corporation contracted with the *Federal Aviation Agency* in 1959 in a program called *SAGE Air Traffic Integration (SATIN)*. The aim of this project was the development of a single, nationwide system for tracking and managing all aircraft in that nation's airspace.[29] This work was, of course, based on the early ATC studies performed by DAVID R. ISRAEL[30] in 1950 and 1951 using Whirlwind.[31] Eventually, but only for a limited time, even one SAGE site[32] became part of the *Federal Aviation Administration (FAA)* air route traffic control center.

Another direct spin-off of SAGE is the *SABRE*[33] system – the first computerized reservation system developed by IBM. Due to an incredible coincidence, BLAIR SMITH, an IBM salesman, was sitting next to CYRUS ROWLETT SMITH, the CEO of *American Airlines* during a business flight in 1953. As they talked, the idea of using computers – back then still electronic brains for most people – for flight reservations etc. came up and IBM jumped at this opportunity to apply the know-how gained during the (still ongoing) development of the AN/FSQ-7 prototypes to a large scale business application. The first incarnation of this revolutionary system became operational in 1964 after some $ 40,000,000 had been spent on research and development.

14.2.5 SAGE in popular culture

Finally, the AN/FSQ-7 had and still has a remarkable impact on popular culture – its maintenance consoles, featured in the settings of more than 60 movies,[34] are still epitomes of giant computers. It is quite remarkable that the first appearance of an AN/FSQ-7 console in a cinema movie was in 1964 in "Santa Claus Conquers the Martians", at a time when all of the documentation regarding AN/FSQ-7 was still classified. At the time of writing this book, the last movie making use of these marvelous consoles was "Future Shock", episode 22, season 1 of "Flash Forward".

It may be safely assumed that SAGE, and especially AN/FSQ-7, inspired the creation of "WOPR", the fantasy "War Operation Plan Response" computer featured in the 1983 movie "WarGames". The final sequences of this movie show a game of *Tic-tac-toe* being played, which causes WOPR to realize that there would be no winner in a global thermonuclear war, thus saving the world. It may just be a coincidence, but A. ZABLUDOWSKY demonstrated a Tic-tac-toe playing program on Whirlwind in 1955...[35]

---

[28] *ATC* for short.
[29] See [MITRE 2008][p. 12].
[30] 1926–02/15/1994
[31] See [ISRAEL 1951] and [ISRAEL 1950].
[32] Great Falls, Malmstrom AFB
[33] Short for *Semi-automated Business Research Environment*.
[34] See http://www.starringthecomputer.com.
[35] See [ZABLUDOWSKY 1955].

15 Epilogue

SAGE, and particularly the development of AN/FSQ-7, was clearly one of the biggest adventures in the history of computing and will always be remembered not only as such, but also as one of the most fertile projects of all time. It also demonstrated the amount of innovation possible on a basically unrestricted budget – a direct result of the Cold War – together with a healthy environment of open recognition of errors – something next to impossible in today's cost driven environments:

> "We built that big memory, and we didn't go to the steering committee to get approval for it. We didn't go up there and say, 'Now, here's what we ought to do, it's going to cost this many million dollars, it's going to take us this long, and you must give us approval for it.' We just had a pocket of money that was for advanced research. We didn't tell anybody what it was for; we didn't have to. Take any one of those developments – whether it was that memory, the Memory Test Computer, or the cathode-ray tubes and the Charactron tubes – if we had had to go through the management stuff that we have to go through now to get $100,000 worth of freedom, we would never have done any of them."[1]

Some institutions established during the developments of Whirlwind and AN/FSQ-7 are still in business today. Lincoln Laboratory, for example, is doing research on communication systems, cyber security and information sciences in general, tactical systems, air traffic control,[2] air and missile defense technology, etc.:

> "[Lincoln] Laboratory did not close down; it entered its second era, one characterized by a significant reduction in activity. Between 1958 and 1960, funding fell by nearly 30%. Yet during this period of uncertainty, it became very clear that much of the work on SAGE was of value to other programs of national interest. The solid-state physics group, for instance, had already achieved an international reputation in its own right. The long-range communications group, originally devoted to SAGE, had embarked on a major effort to export the feasibility of using passive satellites for communications.[3]

---

[1] See [TROPP et al. 1983][pp. 378 f.].
[2] See [FIELDING 1994].
[3] See [GROMETSTEIN 2011][p. xiii].

Figure 15.1: Console of the AN/FSQ-32

Even AN/FSQ-7's ill-fated successor, the *AN/FSQ-7A* or *AN/FSQ-32*, of which only two prototypes were actually built, had some impact on the field of computer science. This machine, affectionately called *Super SAGE*, featured a word length of 48 bits, was fully transistorized, liquid cooled, and clocked at a remarkable 6.4 MHz. Figure 15.1 shows the central maintenance console of one of the two AN/FSQ-32 systems which proved to be extremely reliable. Eventually, one system was installed at an IBM research center while the other computer found its home at SDC. This latter machine would become part of early experiments in packet switched networks in October 1965 showing their practicability and further paving the way to today's ubiquitous computer networks.[4]

Thanks to many individuals, a wealth of documents regarding Whirlwind and AN/FSQ-7 has been preserved.[5] Unfortunately, much of AN/FSQ-7's hardware has been lost – although some museums have parts of these marvelous machines on display, no complete installation has survived. The situation regarding actual software, ranging from diagnostics to the DCAs and CCAs is even worse as nothing has surfaced to date.

As the heritage of AN/FSQ-7 still shapes our world in ways unforeseen in the 1950s, it is to be hoped that future generations will realize the importance and value of such technological artifacts and preserve them as vital parts of the history of our culture.

---

[4]See http://www.packet.cc/internet.html, retrieved 04/04/2014.
[5]See http://www.bitsavers.org/pdf/ibm/sage/.

A Whirlwind instruction set

The following list of Whirlwind instructions is a transcription of [MUHLE 1958] [pp. 14 ff.]:

Operation	Opcode	Execution time in μs	Description
si pqr	00000	30	Select in/out unit/stop: Stop any in/out unit that may be running and select a particular in/out unit and start it or stop the computer. p, q and r select the unit and mode of operation.
—	00001	N/A	Illegal instruction – its use will result in a check alarm.
bi x	00010	≈ 8 ms	Block transfer in: Transfer a block of n words or characters from an in/out unit to core memory, where x is the start address in core memory and the AC contains n.
rd	00011	15	Read: Transfer a word from Input/Output Register (IOR) to AC and then clear IOR.
bo x	00100	≈ 8 ms	Block transfer out: Transfer a block of n words from core memory to an in/out unit. x contains the start address of the block in memory and the AC contains the number of words to be transferred.
rc	00101	22	Record: Transfer the contents of the AC to an in/out unit via the IOR.
sd x	00110	22	Sum digits: The sum of the original contents of digit i of the AC and the contents of digit i in x is stored in digit i of the AC (exclusive or).
cf pqr	00111	15	Change fields: Memory management (see [MUHLE 1958][p. 16]).
ts x	01000	22	Transfer to storage: Transfer the contents of the AC to memory location x.
td x	01001	29	Transfer digits: Transfer the last 11 digits of the AC to the last 11 digits of memory location x.
ta x	01010	29	Transfer address: Like td but using AR instead of AC. This instruction is normally used after an sp or cp instruction (subroutine handling).

Operation	Opcode	Execution time in μs	Description
ck x	01011	22	Check: Compare the contents of the AC with the contents of memory location x. Depending on the setting of the *Program Check Alarm on Special Mode* switch on the console, the instruction behaves differently: In *normal mode*, the next instruction will be executed if the values are identical, otherwise a *check register alarm* will be initiated. In *special mode*, no such alarm will be generated but the next instruction will be skipped if the values are non-equal.
ab x	01100	29	Add BR: Add the contents of the BR to the contents of memory location x and store the result in the AC and in memory location x. The contents of x remain in AR. On overflow an arithmetic check alarm is given. Special Add Memory (SAM) is cleared.
ex x	01101	29	Exchange: The contents of the AC are exchanged with those of memory location x. ex 0 will clear the AC without clearing BR.
cp x	01110	15	Conditional program: If the contents of the AC are negative, proceed as in sp. Otherwise, proceed to the next instruction, clear AR and place the address of this next instruction into the last 11 digits of the AR.
sp x	01111	15	Sub-program: Store the contents of the PC (i.e. the address from which the sp instruction was read) plus one into AR and load the PC with x, thus transferring control to the instruction in memory location x.
ca x	10000	22	Clear and add: Clear AC and BR, then obtain the contents of SAM (+1, 0, or -1) times 2^{-15} and add the contents of memory location x, storing the result in the AC. The contents of location x appear in AR, SAM is cleared. On overflow an arithmetic check alarm is given.
cs x	10001	22	Clear and subtract: Like ca but subtracting the contents of memory location x.
ad x	10010	22	Add the contents of memory location x to the contents of the AC, storing the result in AC. The contents of location x appear in AR. An overflow will give an arithmetic check alarm.
su x	10011	22	Subtract: Like ad but subtracting the contents of memory location x from AC.

A Whirlwind instruction set

Operation	Opcode	Execution time in μs	Description
cm x	10100	22	Clear and add magnitude: Clear AC and BR, then obtain contents of SAM (+1, 0, or -1) times 2^{-15} and add magnitude of contents of memory location x, storing the result in AC. The magnitude of the contents of memory location x appears in AR, SAM is cleared. In case of an overflow, an arithmetic check alarm is given.
sa x	10101	26	Special add: Add the contents of memory location x to the contents of AC, storing the fractional result in AC and retaining any overflow (including sign) in SAM for use with a following ca, cs or cm instruction. (Be careful not to destroy the contents of SAM between sa and one of these instructions.)
ao x	10110	29	Add one: Add one times 2^{-15} to the contents of memory location x, storing the result in this location and in the AC. The original contents of location x appear in AR, SAM is cleared. An overflow will give an arithmetic check alarm.
dm x	10111	22	Difference of magnitudes: Subtract the magnitude of the contents of memory location x from the magnitude of the contents of AC, leaving the result in AC. The magnitude of the contents of location x appears in AR, SAM is cleared, BR will contain the initial contents of the AC. If the absolute value of the contents of AC and the memory location x are identical, the result will be -0.
mr x	11000	34–41	Multiply and roundoff: Multiply the contents of AC by the contents of the memory location x. Roundoff result from BR to 15 significant binary digits and store it in AC. If bit 0 of BR is one, roundoff is 2^{-15}, otherwise zero. Clear BR. The magnitude of the contents of memory location x appears in AR, SAM is cleared. The sign of the AC is determined by the sign of the product.
mh x	11001	34–41	Multiply and hold: Multiply the contents of AC by the contents of memory location x. Retain the full product in AC and in the first 15 digit positions of BR, the last bit of BR will be cleared. The magnitude of the contents of memory location x appears in AR, SAM is cleared. The result is a double register product.

Operation	Opcode	Execution time in μs	Description
dv x	11010	71	Divide: Divide contents of AC by the contents of memory location x, leaving 16 bits of the quotient in BR and ±0 in AC according to the sign of the quotient. The instruction slr following the dv operation will roundoff the quotient to 15 bits and store it in AC. See [MUHLE 1958][p. 21] for detailed information on overflow conditions and handling.
slr n	11011-0	15+.8n	Shift left and roundoff: Shift fractional contents of AC (except sign digit) and BR n places to the left. The positive integer value n is treated modulo 32. Roundoff the result to 15 bits and store it in AC. Clear BR. Negative numbers are complemented before the shift and after the roundoff; hence, ones appear in the digit places made vacant by the shift of a negative number. Bit six of the instruction must be zero to distinguish slr from slh. slr 0 simply causes roundoff and clears BR. In case of an overflow, an arithmetic check alarm is given.
slh n	11011-1	15+.8n	Shift left and hold: Shift the contents of AC (except sign) and BR n places to the left (the positive integer n is treated modulo 32). Digits shifted left out of AC bit one are lost. (Shifting left n places is equivalent to multiplying by 2^n, with the result reduced modulo 1.) Leave the final product in AC and BR. Do not roundoff or clear BR. Negative numbers are complemented in AC before and after the shift; hence, ones appear in the digit places made vacant by the shift of a negative number.
srr n	11100-0	15+.8n	Shift right and roundoff: Shift contents of AC (except sign digit) and BR to the right n places (the positive integer n is treated modulo 32); digits shifted right out of bit 15 of BR are lost. (Shifting right n places is equivalent to multiplying by 2^{-n}.) Roundoff the result to 15 bits and store it in AC. Clear BR. Negative numbers are complemented before shift and after the roundoff; hence, ones appear in the digit places made vacant by the shift of a negative number. srr 0 simply causes roundoff and clears BR. SAM is cleared. Roundoff may cause an overflow with a consequent arithmetic check alarm.

A Whirlwind instruction set

Operation	Opcode	Execution time in μs	Description
srh n	11100-1	15+.8n	Shift right and hold: Shift contents of AC (except sign digit) and BR n places to the right (the positive integer n is treated modulo 32); digits shifted right out of bit 15 of BR are lost. (Shifting right n places is equivalent to multiplying by 2^{-n}.) Do not round-off the result of clear BR. The result is stored in AC and BR. Negative numbers are complemented in AC before and after the shift; hence, ones appear in the digit places made vacant by the shift of a negative number. SAM is cleared.
sf x	11101	30–76	Scale factor: Multiply the contents of AC and BR by 2 often enough to make the positive magnitude of the product equal to or greater than 1/2. Leave the final product in AC and BR. Store the number of multiplications in AR and in the last 11 bits of the memory location x (the first five bits of this word are left undisturbed). If all the digits in BR are zero and AC contains ±0, this instruction leaves AC and BR undisturbed and stores the number 33 times 2^{-15} in AR and the last 11 bits of memory location x. Negative numbers are complemented in AC before and after the multiplication (by shifting); hence, ones appear in the digit places made vacant by the shift. SAM is cleared.
clc n	11110-0	15+.8n	Cycle left and clear (BR): Shift the full contents of AC (including the sign bit) and BR n places to the left (the positive integer n is treated modulo 32); digits shifted left out of bit 0 of the AC are carried around into bit 15 of BR so that no digits are lost. No roundoff. Clear BR. No complementing before or after the shift takes place. The digit finally shifted into the sign digit position determines whether the result is to be considered a positive or negative quantity. clc 0 simply clears BR without affecting AC.
clh n	11110-1	15+.8n	Cycle left and hold: Shift the full contents of AC (including the sign bit) and BR n places to the left (the positive integer n is treated modulo 32); digits shifted left out of bit 0 of the AC are carried around into bit 15 of BR so that no digits are lost. No complementing takes place; the result is kept in AC and BR. clh 0 does nothing.

Operation	Opcode	Execution time in μs	Description
md x	1111	22	Multiply digits with no roundoff: Bitwise AND of the contents of AC and memory location x, the result is stored in AC, AR contains the complement of the final contents of AC.

B Programming cards

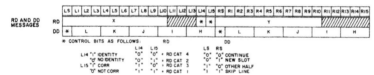
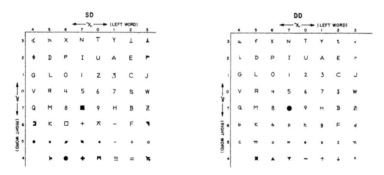

Figure B.1: Programming Data Card, front side (courtesy of MIKE LOEWEN)

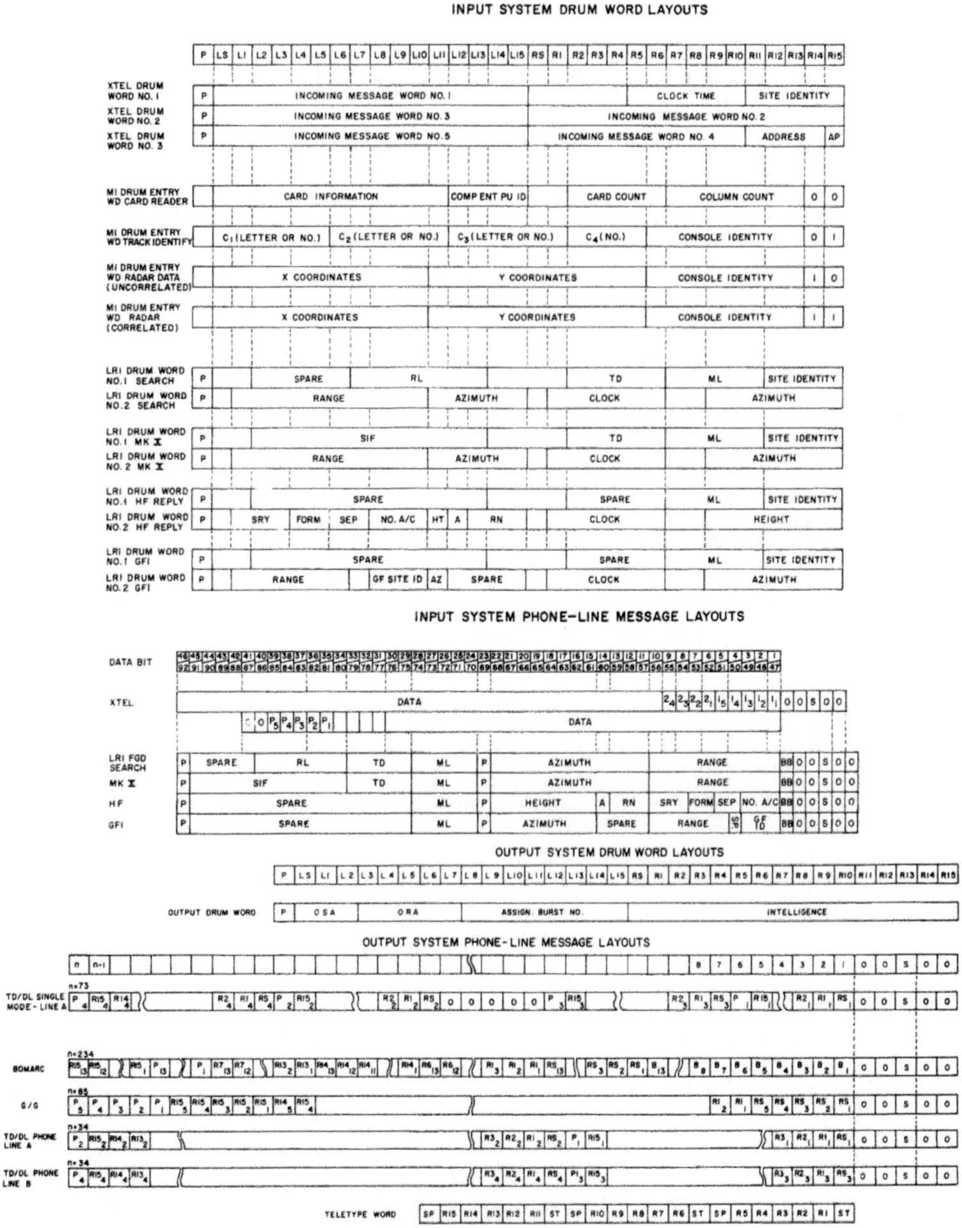

Figure B.2: Programming Data Card, back side (courtesy of MIKE LOEWEN)

B Programming cards

SELECT		Marginal Ck. Exc. On	20	Card Punch 1 & 2 73–74
Card Reader	01	Sense Sw's 1–4 On	21–24	Lock Address Ctr. 75
Card Punch	02	Printer 1 & 2	31–32	Reset Scan Ctr. 76
Printer	03	Output Alarm On	33	Step Scan Ctr. 77
IO Register	04	Output Frame	34, 36–40	
Mech. Clock	05	Display Camera On	35	**SELECT DRUMS**
Man. Input Matrix	06			AM 1–12 02–15
Warning Lights	10	**OPERATE**		IC Other 16
Tapes 1–6	11–16	Cond. Lts. 1–4	01–04	DD Test 17
Burst Time Ctr's	21	Area Disr. 1 & 2	17 & 20	Spares 1 & 2 20–12
		Mar. Ck. Start	21	MI 22–23
SENSE		Mar. Ck. Stop	22	Crosstell 24–25
Cond. Lts. 1–4 On	01–04	Clear IO Interlock	27	IC Own 26
Tapes Not Prep.	10	Initiate Display		DD 27
IO Not Ready	11	Camera 1 & 2	31–32	OB Odd 30
L. Overflow On	12	Initiate DD 1 & 2	35–36	OB Even 31
R. Overflow On	13	Printer 1–10	51–62	LRI 32–33
IO Interlock On	14	Set Prepared	67	GFI 34–37
Mem. Parity Error	15	Backspace Tape	70	Crosstell Marker 40
Drum Parity Error	16	Rewind Tape	71	TD 1–6 41–46
Tape Parity Error	17	Write EOF	72	SD Test 47
				RD 1–9 60–60

Table B.1: Octal index interval codes for various instruction classes

OCTAL OPERATION CODES

MISC	MULT	BRANCH
HLT 000	*MUL 250	BPX 51-
*ETR 004	*TMU 254	BSN 52-
PER 01-	*DVD 260	BFZ 540
CSW 020	*TDV 264	BFM 544
SLR 024#		BLM 550
*LDB 030		BRM 554
*CMM 040	**STORE**	
*CDM 041#		
*CMR 042	*FST 324	**IN/OUT**
*CDR 043#	*LST 330	
*CML 044	*RST 334	*LDC 600
*CDL 045#	*STA 349	*SDR 61-
*CMF 046	*AOR 344#	*SEL 62-
*CDF 047#	*ECH 350	*RDS 670
*TOB 050	*DEP 360	*WRT 674
*TTB 054		

ADD	SHIFT	RESET
*CAD 100	DSL 400	XIN 754
*ADD 104#	DSR 404	XAC 764
*TAD 110#	ASL 420	ADX 770
*ADB 114#	ASR 424	
*CSU 130	LSR 440	
*SUB 134#	RSR 444	
*TSU 140#	DCL 460	
*CAM 160	FCL 470	
*DIM 164		
*CAC 170		

* Indexable
Has 17-bit option
\# Can cause overflow

Table B.2: AN/FSQ-7 (XD-1) Programming Data for Central Computer (Card #1), 1 April 1958 (includes changes scheduled through October 1958)

Bibliography

[ADAMS et al. 1951] CHARLES W. ADAMS, P. A. FOX, *Notes on the Logical Design of Digital Computers and on Special Coding Techniques, Spring Term – 1951*, Electronic Computer Division, Servomechanics Laboratory, Massachusetts Institute of Technology

[ADAMS 1954] CHARLES W. ADAMS, *The M.I.T. Systems of Automatic Coding: Comprehensive, Summer Session, and Algebraic*, Report R-233, Digital Computer Laboratory, Massachusetts Institute of Technology, July 12, 1954

[ADAMS et al. 1954] CHARLES W. ADAMS, S. GILL, D. COMBALIC, *Digital Computers – Advanced Coding Techniques*, Massachusetts Institute of Technology, Summer Session 1954

[ADC SDC] *SAGE – ADC Computer Programming & System Training Office – and the System Development Corporation*, Air Defense Command, Computer Programming and System Training Office, 24 April 1961

[Armed Services Press 1979] Armed Services Press, *the Syracuse area salutes HANCOCK FIELD*, 1979

[ASTIA 401 412] Armed Services Technical Information Agency (ASTIA), *Identification of Problem Areas in the Weapons Director Function Through Critical Incidents*, Tech Memo, AD 401 412, Apr. 15, 1963

[ASTRAHAN et al. 1983] MORTON M. ASTRAHAN, JOHN F. JACOBS, "History of the Design of the SAGE Computer – The AN/FSQ-7", in *Annals of the History of Computing*, Volume 5, Number 4, October 1983, pp. 340–349

[BAGLEY 1954] PHILIP R. BAGLEY, *Memory Test Computer: Programming Manual*, Memorandum M-2527-1, Division 6 – Lincoln Laboratory, Massachusetts Institute of Technology, November 25, 1953, Revised April 23, 1954

[BAGLEY et al. 1954] PHILIP R. BAGLEY, H. H. DENMAN, *Programming for In-Out Units*, Memorandum M-1623-2, Digital Computer Laboratory, Massachusetts Institute of Technology, September 1954

[BARRY 1949] F. N. BARRY, "Receiving Tubes", in [BLACKBURN 1949], pp. 517–613

[BELL et al. 1949] P. R. BELL, G. D. FORBES, E. F. MACNICHOL, Jr., "Storage Tubes", in [CHANCE et al. 1949][pp. 707–729]

[BELL 1983] GORDON BELL, "Field Trip to North Bay", in *The Computer Museum Report*, Spring 1983, pp. 13–14

[BENINGTON 1983] HERBERT D. BENINGTON, "Production of Large Computer Programs", in *Annals of the History of Computing*, Volume 5, Number 4, October 1983, pp. 350–361

[Biweekly Report 3674] *Biweekly Report for 27 May 1955*, Division 6 – Lincoln Laboratory, Massachusetts Institute of Technology, 27 May 1955

[BLACKBURN 1949] JOHN F. BLACKBURN, *Components Handbook*, MIT Radiation Laboratory Series, Volume 17, McGraw-Hill, 1949

[BLOCH 1951] RICHARD M. BLOCH, "The Raytheon Electronic Digital Computer", in *Proceedings of a Second Symposium on Large-Scale Digital Calculating Machinery*, Harvard University Press, 1951, pp. 50–64

[BLODGETT 1955] E. O. BLODGETT, *Tape Controlled Typewriter*, United States Patent 2700446, Jan. 25, 1955

[BOSLAUGH 2003] DAVID L. BOSLAUGH, *When Computers Went to Sea – The Digitization of the United States Navy*, IEEE Computer Society, 2003

[BOYD 1953] HAL BOYD, *High-Speed (5965) Flip-Flop*, Engineering Note E-526, Digital Computer Laboratory, Massachusetts Institute of Technology, February 24, 1953

[BRIGGS 1988] BRUCE BRIGGS, *The Shield of Faith: a Chronicle of Strategic Defense From Zeppelins to Star Wars*, Simon and Schuster New York, 1988

[BROWN 1953] DAVID R. BROWN, *Magnetic Materials for High-Speed Pulse Circuits*, Engineering Note E-530, Digital Computer Laboratory, Massachusetts Institute of Technology, February 27, 1953

[BRUDERER 2012] HERBERT BRUDERER, *Konrad Zuse und die Schweiz – Wer hat den Computer erfunden?*, Oldenbourg Verlag, 2012

[BUCK et al. 1953] D. A. BUCK, W. I. FRANK, *Nondestructive Sensing of Magnetic Cores*, Engineering Note E-454-1, Digital Computer Laboratory, Massachusetts Institute of Technology, March 24, 1952, revised March 24, 1953

[BURMAN 1972] BRUCE BURMAN, "Vacuum tubes yield sockets to hybrid JFET devices", in *Electronics*, April 1972

[CARR 1952] JOHN W. CARR III, *Automatic Assembly of Programs*, Memorandum M-1445, Digital Computer Library, Massachusetts Institute of Technology, April 23, 1952

[CEM 55-19] *CEM Maintenance Operating Instruction No. 55-19*, Headquarters 20th Air Division (ADC), Fort Lee AFS, Virginia, 1 June 1970

[CHANCE et al. 1949] BRITTON CHANCE, VERNON HUGHES, EDWARD F. MACNICHOL, DAVID SAYRE, FREDERIC C. WILLIAMS (eds.), *Waveforms*, MIT Radiation Laboratory Series, Volume 19, McGraw-Hill, 1949

[Chicago Air Defense Sector] *FACT SHEET*, Headquarters CHICAGO AIR DEFENSE SECTOR, United States Air Force, Truax Field, Madison 7, Wisconsin

[CHILDRESS 1956] JAMES D. CHILDRESS, *Geometry of Magnetic Memory Elements*, Memorandum 6M-4089, Division 6 – Lincoln Laboratory, Massachusetts Institute of Technology, January 18, 1956

[CLARK 1954] W. A. CLARK, *The Multi-Sequence Program Concept*, Memorandum 6M-3144, Division 6 – Lincoln Laboratory, Massachusetts Institute of Technology, 5 November 1954

[COHEN 2000] BERNARD COHEN, *Howard Aiken and the Dawn of the Computer Age*, in [ROJAS et al. 2000][pp. 107–120]

[Commercial Controls] Commercial Controls Corporation, *Flexowriter Model FG*, sales brochure

[CORRELL 2005] JOHN T. CORRELL, "How the Air Force Got the ICBM", in *Air Force Magazine – Online Journal of the Air Force Association*, http://www.airforcemag.com/MagazineArchive/Pages/2005/July%202005/0705icbm.aspx, retrieved 04/03/2014

[CRAWFORD 1939] PERRY ORSON CRAWFORD, Jr., *Automatic Control by Arithmetical Operations*, Massachusetts Institute of Technology, 1939 (1942)

[DAVISON 1977] Maj. DON DAVISON, "25th Region is at forefront of air defense change", in *Skywatch*, 25th NORAD Region/Air Division, September 1977, pp. 10–11

[SKYWATCH] Lt. Col. DON DAVISON, "Clyde's double life reaps AF $$", in *SKYWATCH*, January 1980, pp. 6–7

[DODD et al. 1950] S. H. DODD, H. KLEMPERER, P. YOUTZ, *M.I.T. Electrostatic Storage Tube*, Report R-183, Electronic Computer Division, Servomechanisms Laboratory, Massachusetts Institute of Technology, April 27, 1950

[DYALL 1948] W. T. DYALL, *A Study of the Persistence Characteristics of Various Cathode Ray Tube Phosphors*, Technical Report No. 56, Research Laboratory of Electronics, Massachusetts Institute of Technology, January 16, 1948

[EARNEST et al. 1998] LES EARNEST, JIM WONG, PAUL EDWARDS, *Vigilance and Vacuum Tubes: The SAGE System 1956–63*, transcript of public lecture, Computer History Museum, May 19, 1998

[ECKERT 1951] J. PRESPER ECKERT, "An Electrostatic Memory System", in *Proceedings of a Second Symposium on Large-Scale Digital Calculating Machinery*, Harvard University Press, 1951, pp. 32–43

[ECKERT 1986] J. PRESPER ECKERT, "ENIAC – The Electronic Numerical Integrator and Computer" in *The Computer Museum Report*, Summer/1986, pp. 3–7

[ECKERT 1997] J. PRESPER ECKERT, Jr., "A Survey of Digital Computer Memory Systems", in *Proceedings of the IEEE*, Vol. 85, No. 1, January 1997, pp. 184–197

[Edmonton Journal 1960] "Radar Will Blanket Canada – Air Defence Near Completion", in *Edmonton Journal*, Nov. 10, 1960, p. 18

[EDWARDS 1997] PAUL N. EDWARDS, *The Closed World, Computers and the Politics of Discourse in Cold War America*, The MIT Press, Cambridge (Massachusetts), London (England), 1997

[EDWARDS 2013] BENJ EDWARDS, "The Never-Before-Told Story of the World's First Computer Art (It's a Sexy Dame)", in *The Atlantic*, Jan. 24

[EVANS 1983/1] CHRISTOPHER EVANS, "Conversation: JAY W. FORRESTER interviewed by CHRISTOPHER EVANS", in *Annals of the History of Computing*, Volume 5, Number 3, July 1983, pp. 297–301

[EVANS 1983/2] CHRISTOPHER EVANS, "Reliability of Components", excerpt from an 1975 interview with JAY W. FORRESTER, in *Annals of the History of Computing*, Volume 5, Number 3, July 1983, pp. 399–401

[EVERETT et al. 1947] ROBERT R. EVERETT, F. E. SWAIN, *Whirlwind I Computer Block Diagrams*, Volume 1 and 2, Report R-127, Servomechanisms Laboratory, Massachusetts Institute of Technology, September 4, 1947

[EVERETT 1951] ROBERT R. EVERETT, "The Whirlwind I Computer", in *Joint Computer Conference AIEE/IRE*, December 10–12, 1951, Philadelphia, Pennsylvania, pp. 70–74

[EVERETT et al. 1983] ROBERT R. EVERETT, CHARLES A. ZRAKET, HERBERT D. BENINGTON, "SAGE – A Data-Processing System for Air Defense", in *Annals of the History of Computing*, Volume 5, Number 3, July 1983, pp. 330–339

[EVERETT 1988] ROBERT R. EVERETT, "Discovering a 'New World' of Computing", in *The Computer Museum Report*, Volume 22, Spring 1988, pp. 14

[FELSBERG 1969] Sgt. DAVID F. FELSBERG, *System Maintenance Control (SMC) Computer Information Folder*, MGE-DC, 8 April 1969

[FIELDING 1994] JOHN C. FIELDING, "Air Traffic Control Development at Lincoln Laboratory", in *The Lincoln Laboratory Journal*, Volume 7, Number 2, 1994, pp. 147–148

[FISCHER 2004] CHARLOTTE FROESE FISCHER, *Douglas Rayner Hartree – His Life in Science and Computing*, World Scientific Publishing Company, 2004

[Flight 1957] "BOMARC, Boeing's Long-range A.A. Missile", in *Flight International*, 24 May 1957, p. 687

[FORGIE 1953] JAMES W. FORGIE, *The WWI Auxiliary Magnetic Drum System*, Engineering Note E-520, Digital Computer Laboratory, Massachusetts Institute of Technology, January 9, 1953

[FORRESTER et al. 1948] JAY W. FORRESTER, HUGH R. BOYD, ROBERT R. EVERETT, HARRIS FAHNESTOCK, ROBERT A. NELSON, *Forecast for Military Systems Using Electronic Digital Computers*, Servomechanisms Laboratory, Massachusetts Institute of Technology, September 17, 1948

[FORRESTER 1950] JAY W. FORRESTER, *Digital Information Storage in Three Dimensions Using Magnetic Cores*, Report R-187, Servomechanisms Laboratory, Massachusetts Institute of Technology, May 16, 1950

[FORRESTER 1951] JAY W. FORRESTER, "The Digital Computation Program at Massachusetts Institute of Technology", in *Proceedings of a Second Symposium on Large-Scale Digital Calculating Machinery*, Harvard University Press, 1951, pp. 44–49

[FORRESTER 1988] JAY W. FORRESTER, "Whirlwind's Success", in *The Computer Museum Report*, Volume 22, Spring 1988, pp. 10–13

[FRIEDMAN 2008] NORMAN FRIEDMAN, *Naval Firepower – Battleship Guns and Gunnery in the Dreadnought Era*, Seaforth Publishing, 2008

[GILMORE 1951] J. T. GILMORE, *Operational Procedure on the Whirlwind Computer*, Memorandum M-1350, Digital Computer Laboratory, Massachusetts Institute of Technology, December 10, 1951

[GLOVER 1955/1] E. B. GLOVER, *Digital Data Transmitter*, Memorandum 6M-3402, Division 6 – Lincoln Laboratory, Massachusetts Institute of Technology, 3 March 1955

[GLOVER 1955/2] E. B. GLOVER, *Digital Data Receiver and Gap Filler Input Receiver*, Memorandum 6M-3403, Division 6 – Lincoln Laboratory, Massachusetts Institute of Technology, 1 April 1955

[GREEN 2010] THOMAS J. GREEN, *Bright Boys*, A K Peters, Ltd., 2010

[GROMETSTEIN 2011] ALAN A. GROMETSTEIN (ed.), *MIT Lincoln Laboratory – Technology in Support of National Technology*, Lincoln Laboratory, Massachusetts Institute of Technology, Lexington, Massachusetts, 2011

[GROSCH 1952] H. R. J. GROSCH, *Signed Ternary Arithmetic*, Memorandum M-1496, Digital Computer Laboratory, Massachusetts Institute of Technology, May 22., 1952

[GUSTAFSON 2000] JOHN GUSTAFSON, *Reconstruction of the Atanasoff-Berry Computer*, in [ROJAS et al. 2000][pp. 91–106]

[HADDAD 1983] JERRIER A. HADDAD, "701 Recollections", in *Annals of the History of Computing*, Volume 5, Number 2, October 1983, pp. 118–124

[HAIGH 2006] THOMAS HAIGH, "'A veritable bucket of facts' origins of the data base management system", in *ACM SIGMOD Record*, Volume 35, Issue 2, June 2007, pp. 33–49

[HARRINGTON et al. 1954] JOHN V. HARRINGTON, PAUL ROSEN, DANIEL A. SPAETH, *Some Results on the Transmission of Pulses over Telephone Lines*, Massachusetts Institute of Technology, Lincoln Laboratory, April 1954

[HARRINGTON 1983] JOHN V. HARRINGTON, "Radar Data Transmission", in *Annals of the History of Computing*, Volume 5, Number 4, October 1983, pp. 370–374

[HEART 1952] F. E. HEART, *Initial Operation of WWI Terminal Equipment with the New In-Out System*, Memorandum M-1551, Digital Computer Laboratory, Massachusetts Institute of Technology, Massachusetts, July 17, 1952

[HITTINGER 1989] WILLIAM C. HITTINGER, "JAN A. RAJCHMAN, 1911–1989", in *Memorial Tributes*, Volume A, National Academy Press, Washington, D. C., 1992

[HOLDEN 2011] H. HOLDEN, *World's only all Fetron radio and the Omega device*, http://www.listenersguide.org.uk/pdf/fetron-valve-replacement.pdf, retrieved 03/23/2014

[HUGHES Aircraft Corporation] HUGHES Aircraft Corporation, *Typotron Direct View Character-Writing Storage Tube*

[HUGHES 1949] VERNON HUGHES, "Electrical Delay Lines", in [CHANCE et al. 1949] [pp. 730–750]

[HUSKEY et al. 1962] HARRY D. HUSKEY, GRANINO A. KORN, *Computer Handbook*, McGraw-Hill Book Company, Inc., 1962

[HUSKEY 2000] HARRY D. HUSKEY, "Hardware Components and Computer Design", in [ROJAS et al. 2000][pp. 69–87]

[IBM AN/FSQ-7] *Theory of Operation of Central Computer System for AN/FSQ-7 Combat Direction Central and AN/FSQ-8 Combat Control Central*, Military Products Division, International Business Machines Corporation, Kingston, New York, 1 February 1959

[IBM BASIC CIRCUITS] *Basic Circuits for AN/FSQ-7 Combat Direction Central*, Military Products Division, International Business Machines Corporation, Kingston, New York, 15 November 1956, revised 1 April 1957

[IBM BOMARC] *The SAGE/BOMARC Air Defense Weapons System – An illustrated explanation of What It Is and How It Works*, Military Products Division, International Business Machines Corporation, N. Y., 1958

[IBM CCS I] *Theory of Operation of Central Computer System for AN/FSQ-7 Combat Direction Central and AN/FSQ-8 Combat Control Central*, Volume I, Military Products Division, International Business Machines Corporation, Kingston, New York, 1 February, 1959

[IBM CCS XD] *Theory of Operation AN/FSQ-7 (XD-1, XD-2) Combat Direction Central, Central Computer System*, Preliminary Manual, Military Products Division, International Business Machines Corporation, Kingston, New York, September 1955

[IBM DRUM] *Theory of Operation of Drum System for AN/FSQ-7 Combat Direction Central and AN/FSQ-8 Combat Control Central,* Volume I, Military Products Division, International Business Machines Corporation, Kingston, New York, 15 September 1958

[IBM DSP 1] *Theory of Operation of Display System for AN/FSQ-7 Combat Direction Central and AN/FSQ-8 Combat Control Central,* Volume I, Military Products Division, International Business Machines Corporation, Kingston, New York, 1 August 1958

[IBM DSP 2] *Theory of Operation of Display System for AN/FSQ-7 Combat Direction Central and AN/FSQ-8 Combat Control Central,* Volume II, Military Products Division, International Business Machines Corporation, Kingston, New York, 1 August 1958

[IBM INPUT] *Theory of Operation of Input System for AN/FSQ-7 Combat Direction Central and AN/FSQ-8 Combat Control Central,* Military Products Division, International Business Machines Corporation, Kingston, New York, 1 December 1958

[IBM INTRO] *Introduction to AN/FSQ-7 Combat Direction Central and AN/FSQ-8 Combat Control Central,* Military Products Division, International Business Machines Corporation, Kingston, New York, 1 January 1959

[IBM MC] *Theory of Operation of Marginal Checking for AN/FSQ-7 Combat Direction Central and AN/FSQ-8 Combat Control Central,* Military Products Division, International Business Machines Corporation, Kingston, New York, 1 December 1958

[IBM MEMORY ANALYSIS] 64^2 *Memory – Analysis of Maintenance Techniques,* Military Products Division, International Business Machines Corporation, Kingston, New York, July, 1958

[IBM OUT] *Theory of Operation of Output System for AN/FSQ-7 Combat Direction Central and AN/FSQ-8 Combat Control Central,* Military Products Division, International Business Machines Corporation, Kingston, New York, 1 December 1958

[IBM PGM] *Theory of Programming for AN/FSQ-7 Combat Direction Central,* Military Products Division, International Business Machines Corporation, Kingston, New York, 15 November 1956

[IBM PGM 1959] *Theory of Programming for AN/FSQ-7 Combat Direction Central and AN/FSQ-8 Combat Control Central,* Military Products Division, International Business Machines Corporation, Kingston, New York, 1 April 1959

[IBM POWER] *Theory of Operation of Power Supply System for AN/FSQ-7 Combat Direction Central and AN/FSQ-8 Combat Control Central,* Military Products Division, International Business Machines Corporation, Kingston, New York, 1 September 1958

[IBM RELAYS] *IBM Customer Engineering Reference Manual, RELAYS,* International Business Machines Corporation, New York, 1950

[IBM SPECIAL CIRCUITS] *Special Circuits for AN/FSQ-7 Combat Direction Central and AN/FSQ-8 Combat Control Central*, Military Products Division, International Business Machines Corporation, Kingston, New York, 1 May 1959

[ISRAEL 1950] DAVID R. ISRAEL, *The Application of a High-Speed Digital Computer to the Present-Day Air Traffic Control System*, 6R-203, Division 6 – Lincoln Laboratory, Massachusetts Institute of Technology, January 15, 1952

[ISRAEL 1951] DAVID R. ISRAEL, *Description of Basic Track-While-Scan and Interception Program*, Memorandum M-1343, Digital Computer Laboratory, Massachusetts Institute of Technology, December 3, 1951

[JACOBS 1986] JOHN F. JACOBS, *The SAGE Air Defense System – a Personal History*, The MITRE Corporation, 1986, second printing 1990

[JOHNSON 2002] STEPHEN B. JOHNSON, *The United States Air Force and the Culture of Innovation*, Air Force History and Museums Program, Washington, D. C., 2002

[Крысенко 1966] Крысенко, Современные Системы ПВО, Военное Издательство Министерства Обороны СССР, Москва, 1966

[LANGE 2006] THOMAS H. LANGE, *Peenemünde – Analyse einer Technologieentwicklung im Dritten Reich*, Reihe *Technikgeschichte in Einzeldarstellungen*, VDI-Verlag, GmbH, Düsseldorf 2006

[LANING et al. 1954] J. H. LANING Jr., N. ZIERLER, *A Program for Translation of Mathematical Equations for Whirlwind I*, Engineering Memorandum E-364, Instrumentation Laboratory, Massachusetts Institute of Technology, January, 1954

[Legacy 2000] *Cold War Needs Assessment – A Legacy Project*, Legacy 98-1754, September 2000

[LOMBARDI 2007] MICHAEL J. LOMBARDI, "Reach for the sky – How the Bomarc missile set the stage for Boeing to demonstrate its talent in systems integration", in *BOEING FRONTIERS*, June 2007, pp. 8–9

[LONE 1952] W. LONE, *Group 61 Subroutine Library*, Memorandum M-1631, Supplement #1, Digital Computer Laboratory, Massachusetts Institute of Technology, October 2, 1952

[LUNDBERG 1955] E. D. LUNDBERG, *SAGE System Meeting 29 August 1955*, Memorandum 6M-3864, Division 6 – Lincoln Laboratory, Massachusetts Institute of Technology, 29 August 1955

[MANN et al. 1954] MARGARET F. MANN, ROBERT R. RATHBONE, JOHN B. BENNETT, *Whirlwind I Operation Logic*, Report R-221, Digital Computer Laboratory, Massachusetts Institute of Technology, 1 May 1954

[MACNICHOL et al. 1949] E. F. MACNICHOL Jr., R. B. WOODBURY, "Blocking oscillators and delay-line pulse generators", in [CHANCE et al. 1949][pp. 205–253]

[MARTIN 1995] DIANNE MARTIN, "ENIAC: The Press Conference That Shook the World", in *IEEE Technology and Society Magazine*, December, 1995

[MIT 1951] *Whirlwind Subroutine Specification*, Digital Computer Laboratory, Massachusetts Institute of Technology, 1951

[MIT 1955] *The Whirlwind I Computer*, April 1955, Massachusetts Institute of Technology

[MIT 1956] *Biweekly Report for Period Ending 4 May 1956*, Memorandum 6M-4322, Division 6 – Lincoln Laboratory, Massachusetts Institute of Technology, 11 May 1956

[MITRE 2008] *The MITRE Corporation – Fifty Years of Service in the Public Interest*, The MITRE Corporation, 2008

[MORRISON 1954] *The Consolidated Test Program – T-3432*, Memorandum M-3058, Division 6 – Lincoln Laboratory, Massachusetts Institute of Technology, 20 September 1954

[MUHLE 1958] CYNTHIA MUHLE, *Whirlwind Programming Manual*, Memorandum 2M-0277, Massachusetts Institute of Technology, Lincoln Laboratory, 31 October 1958

[MURPHY 1972] HOWARD R. MURPHY, *The Early History of the MITRE Corporation – Its Background, Inception, and First Five Years*, Volume One of Two Volumes, M72-110, State University College Oneonta, New York, Draft, June 30, 1972, Reprinted May 12, 1976

[NYADS 1960] *NYADS Yearbook*, 1960

[O'BRIEN 2010] FRANK O'BRIEN, *The Apollo Guidance Computer – Architecture and Operation*, Springer, Praxis Publishing, 2010

[OGLETREE et al. 1957] W. A. OGLETREE, H. W. TAYLOR, E. W. VEITCH, J. WYLEN, *AN/FST-2 radar-processing equipment for SAGE*, in IRE-ACM-AIEE '57 (Eastern) Papers and discussions, presented at the December 9–13, 1957, eastern joint computer conference: Computers with deadlines to meet, pp. 156–160

[ORNSTEIN 2002] SEVER M. ORNSTEIN, *Computing in the Middle Ages – A View from the Trenches 1955–1983*, 1stBooks, 2002

[PRESS 1996] LARRY PRESS, "Seeding networks: the federal role", in *Communications of the ACM*, Volume 39, Issue 10, Oct. 1996, pp. 11–18

[PWSR 1951] *Project Whirlwind – Summary Report No. 28, Fourth Quarter 1951*, Digital Computer Laboratory, Massachusetts Institute of Technology

[PWSR 1953] *Project Whirlwind – Summary Report No. 35, Third Quarter 1953*, Digital Computer Laboratory, Massachusetts Institute of Technology

[Rajchman 1957] Jan A. Rajchman, *Magnetic System*, United States Patent 2792563, May 14, 1957

[Rathbone 1951] R. R. Rathbone, *Whirlwind 1*, Report R-209, Digital Computer Laboratory, Massachusetts Institute of Technology, 15 August 1951

[RCA 5965] *5965 Medium-Mu Twin Triode 5965*, RCA, Tube Division, June 14, 1954

[Redlands 1960] Redlands Daily Facts, 15 September 1960

[Redmond et al. 1975] Kent C. Redmond, Thomas M. Smith, *Project Whirlwind Case History*, Manuskript Edition, Reproduced by The MITRE Corporation, Bedford, MA, November 1975

[Redmond et al. 1980] Kent C. Redmond, Thomas M. Smith, *Project WHIRLWIND – The History of a Pioneer Computer*, Digital Press, 1980

[Redmond et al. 2000] Kent C. Redmond, Thomas M. Smith, *From Whirlwind to MITRE – The R&D Story of the SAGE Air Defense Computer*, The MIT Press, 2000

[Reilly 2003] Edwin D. Reilly, *Milestones in Computer Science and Information Technology*, Greenwood Press, 2003

[Rhodes 2005] Richard Rhodes, *Dark Sun – the Making of the Hydrogen Bomb*, Simon & Schuster Paperbacks edition, 2005

[Rich 1950] Edwin S. Rich, *Computer Experience in Extending Tube Life*, Report R-184, Electronic Computer Division, Servomechanisms Laboratory, Massachusetts Institute of Technology, April 27, 1950

[Rich 1951] Edwin S. Rich, *Operation of Magnetic Drums with WWI*, Memorandum M-1358, Digital Computer Laboratory, Massachusetts Institute of Technology, December 27, 1951

[Rochester 1983] Nathaniel Rochester, "The 701 Project as Seen by its Chief Architect", in *Annals of the History of Computing*, Volume 5, Number 4, October 1983, pp. 115–117

[Rojas et al. 2000] Raúl Rojas, Ulf Hashagen, *The First Computers – History and Architectures*, MIT Press, 2000

[Saxenian 1951] Hrand Saxenian, *Programming for Whirlwind I*, Report R-196, Electronic Computer Division, Servomechanism Laboratory, Massachusetts Institute of Technology, June 11, 1951

[Sayre 1949] David Sayre, "Generation of fast waveforms", in [Chance et al. 1949] [pp. 159–204]

[Shortell 1963] Albert V. Shortell, Jr., *Whirlwind I Moving, Reassembly and Checkout Progress Report*, Wolf Research and Development Corporation, July 1963

[SHORTELL 1964] ALBERT V. SHORTELL, Jr., *Whirlwind I Checkout Progress Report*, Wolf Research and Development Corporation, January 1964

[SHULMAN 1959] HARRY G. SHULMAN, "Air Force SAGE Center At Topsham Joins Nationwide Defense Net Friday", in *Portland Sunday Telegram*, 2/26/1959

[SMOTHERMAN 1989] M. SMOTHERMAN, "A sequencing-based taxonomy of I/O systems and review of historical machines", in *ACM SIGARCH Computer Architecture News*, Volume 17, Issue 5, Sep. 1989, pp. 5–15

[STIFLER et al. 1950] W. W. STIFLER, jr. (ed.), C. B. TOMPKINS, J. H. WAKELIN (superv.), *High-Speed Computing Devices*, McGraw-Hill Book Company, Inc., 1950

[STUART-WILLIAMS 1962] RAYMOND STUART-WILLIAMS, "Magnetic-Core Storage and Switching Techniques", in [HUSKEY et al. 1962][pp. 12-41–12-106]

[SUMNER 1950] GEORGE C. SUMNER, *Marginal Checking: Preventive Maintenance for Electronic Equipment*, Report R-185, Electronic Computer Division, Servomechanism Laboratory, Massachusetts Institute of Technology, April 27, 1950

[SVOBODA 1948] ANTONIN SVOBODA, *Computing Mechanisms and Linkages*, MIT Radiation Laboratory Series, Volume 27, McGraw-Hill Book Company, Inc., 1948

[TAYLOR et al. 1952] N. H. TAYLOR, W. A. HOSIER, *Whirlwind II Meeting of July 15, 1952*, Memorandum M-1562, Digital Computer Laboratory, Massachusetts Institute of Technology, July 24, 1952

[TAYLOR 1953] N. H. TAYLOR, *Rudiments of Good Circuit Design*, Report R-224, Digital Computer Laboratory, Massachusetts Institute of Technology, May 19, 1953

[TCM 83] *The Computer Museum Report*, The Computer Museum, One Iron Way, Marlboro, Massachusetts 01752, Winter 1983

[Teledyne 1972] Teledyne Semiconductor, "Fetron – Der echte Röhrenersatz", in *Halbleiter Mitteilungen*, 6/1, Oktober 1972

[Teledyne 1973] Teledyne Semiconductor, *FETRON, solid state vacuum tube replacement*, June 1973

[The Command Post 1958] *The Command Post*, Headquarters 26th Air Division (Defense), special edition, Vol. 1, No. 11, July, 1958

[The New York Times 2001] "Preserving a Hair-Raising Relic of the Cold War", in *The New York Times*, August 28, 2001

[The Tech] *The Tech – Newspaper of the Undergraduates of the Massachusetts Institute of Technology*, Cambridge, Massachusetts, Friday, April 29, 1960

[TOMAYKO 1985] JAMES E. TOMAYKO, "Helmut Hoelzer's Fully Electronic Analog Computer", in *Annals of the History of Computing*, Volume 7, Number 3, July 1985, pp. 227–240

[Tomayko 2000] James E. Tomayko, *Computers Take Flight – a History of NASA's Pioneering Digital Fly-By-Wire Project*, NASA SP-2000-4224, 2000

[Tropp et al. 1983] Henry S. Tropp, Herbert D. Benington, Robert Bright, Robert P. Crago, Robert R. Everett, Jay W. Forrester, John V. Harrington, John F. Jacobs, Albert R. Shiely, Norman H. Taylor, C. Robert Wieser, "A Perspective on SAGE: Discussion", in *Annals of the History of Computing*, Volume 5, Number 4, October 1983, pp. 375–398

[Ulmann 2013] Bernd Ulmann, *Analog Computing*, Oldenbourg-Verlag, 2013

[Valley 1948] George E. Valley, Henry Wallman, *Vacuum Tube Amplifiers*, MIT Radiation Laboratory Series, Volume 18, McGraw-Hill, 1948

[Valvo 1965] *Ringkerne aus Ferroxcube 6 in Speichermatrizen und -blöcken*, Valvo GmbH, April 1965

[Vance et al. 1957] P. R. Vance, L. G. Dooley, C. E. Diss, "Operation of the SAGE duplex computers", in *Proceedings of the Eastern Computer Conference*, IRE-ACM-AIEE '57, pp. 160–163

[Van der Spiegel et al. 2000] Jan Van der Spiegel, Jamed F. Tau, Titiimaea F. Ala'ilima, Lin Ping Ang, *The ENIAC: History, Operation and Reconstruction in VLSI*, in [Rojas et al. 2000][pp. 121–178]

[Viehe 1961] Frederick W. Viehe, *Memory Transformer*, United States Patent US2992414, July 11, 1961

[Viehe 1968] Frederick W. Viehe, *Memory Structure Having Cores Comprising Magnetic Particles Suspended in a Dielectric Medium*, United States Patent 3366940, Jan. 30, 1968

[Wainstein et al. 1975] L. Wainstein, C. D. Cremeans, J. K. Moriarty, J. Ponturo, *The Evolution of U.S. Strategic Command and Control and Warning 1945–1972*, Study S-467, Institute for Defense Analyses, International and Social Studies Division, June 1975

[Whirlwind Programming Notes] Untitled, Whirlwind Programming Notes, https://archive.org/details/bitsavers_mitwhirlwis_1656180, retrieved 12/02/2013

[Wieser 1983] C. Robert Wieser, "The Cape Cod System", in *Annals of the History of Computing*, Volume 5, Number 4, October 1983, pp. 362–369

[Wieser 1988] C. Robert Wieser, "From World War II Radar Systems to SAGE", in *The Computer Museum Report*, Volume 22, Spring 1988, pp. 15–16

[Wildes et al. 1986] Karl L. Wildes, Nilo A. Lindgren, *A Century of Electrical Engineering and Computer Science at MIT, 1882 – 1982*, © 1985 by The Massachusetts Institute of Technology, Second printing, 1986

[WILKES et al. 1951] MAURICE V. WILKES, DAVID J. WHEELER, STANLEY GILL, *The Preparation of Programs for an Electronic Digital Computer – With special reference to the EDSAC and the use of a library of subroutines*, Addison-Wesley Press, Inc., 1951

[WILLIAMS 1962] FREDERIC C. WILLIAMS, "WILLIAMS-Tube Memory System", in [HUSKEY et al. 1962][pp. 12-34–12-41]

[YOUNG 1954] GUY A. YOUNG, *Increased Facilities for Visual Display in the WWI Input-Output System*, Memorandum M-2728, Division 7 – Lincoln Laboratory, Massachusetts Institute of Technology, March 17, 1954

[ZABLUDOWSKY 1955] A. ZABLUDOWSKY, *Tic Tac Toe Playing Routine*, Digital Computer Laboratory, Massachusetts Institute of Technology, DCL-105, 4 October 1955

[ZEMANEK 1977] H. ZEMANEK, "EDV – Glaubensbekenntnis oder Hilfsmittel?", in *Elektronische Rechenanlagen*, 19. Jahrgang, 1977, Heft 1, p. 5–8

[ZIEGLER 1957] H. L. ZIEGLER, *The MTC Service Manual*, Division 6 – Lincoln Laboratory, Massachusetts Institute of Technology, 15 April 1957

[ZUSE 1993] KONRAD ZUSE, *Der Computer – Mein Lebenswerk*, Springer Verlag, 3., unveränderte Auflage, 1993

Acronyms

AC Accumulator. 27, 30, 31, 63, 64, 131–134, 192, 194, 197, 204, 223–228

AD Air Defense. 2

ADSEC Air Defense Evaluation Group. 13, 14

AE Arithmetic Element. 19, 119, 125, 132, 202

AFB Air Force Base. 77–79, 115, 125, 209, 212, 217, 220

AFCRC Air Force Cambridge Research Center. 16

AFS Air Force Station. 77, 78

AI Airborne Intercept. 16

AM Auxiliary Memory. 139, 140

APC Angular Position Counter. 49, 50

AR A register. 27, 30, 31, 64, 131–134, 195, 223–225, 227, 228

ASCA Aircraft Stability and Control Analyzer. 13

ASTIA Armed Services Technical Information Agency. 233

ATC Air Traffic Control. xii, 173, 220

BO1 BOMARC 1. 160

BO2 BOMARC 2. 160

BR B register. 27, 30, 31, 63, 131, 194, 197, 204, 224–227

CB Circuit Breaker. 87

CC Control Center. 76–78, 82, 83, 139, 140, 149, 150, 156–158, 160, 165, 170, 176, 183, 203, 208, 209, 211, 212, 216

CCA Combat Center Active. 83, 85, 209, 212, 213, 219, 222

CD Computer-Drum. 140, 146

CEP Computer Entry Punch. 181

CFB Canadian Forces Base. 78

CRL Air Force Cambridge Research Laboratory. 54

CRT Cathode Ray Tube. 56

DAC Digital to Analog Converter. 57, 113

DC Direction Center. 56, 71, 75–79, 81–83, 125, 139, 149, 150, 153, 154, 156–158, 160, 165, 170, 173, 176, 183, 203, 208, 209, 211, 212, 216, 218

DCA Direction Center Active. 83, 85, 209, 212–214, 219, 222

DCL Digital Computer Laboratory. 13, 15, 23, 37

DD Digital Display. 139, 140, 165, 167, 174–176, 213, 231

DDGE Digital Display Generator Element. 87, 175

DDR Digital Data Receiver. 55

DDT Digital Data Transmitter. 54, 55

DFD Drum Field Driver. 141, 142

DRD Drum Read Driver. 142, 143

DRR Digital Radar Relay. 16, 17

ECCMT Electronic Counter Counter Measures Technician. 216

FAA Federal Aviation Administration. 220

FD Frequency Diversity. 216

G/A-FD Ground-to-Air Frequency-Division. 159, 161

G/A-TD Ground-to-Air Time-Division. 159

G/G Ground-to-Ground. 160

GCI Ground Control Intercept. 13

GF Gap Filler. 71

GFI Gap-Filler Input. xi, 82, 86, 87, 139, 140, 146, 147, 149, 150, 153–155, 157, 173, 181, 212, 231

GFIR Gap Filler Input Receiver. 55

IAP International Airport. 78

2 Acronyms

IC Intercommunication. 139, 140

ICBM Intercontinental Ballistic Missile. 215, 216

IFF Identification Friend or Foe. 54, 74, 150

IOR Input/Output Register. 223

LRI Long-Range Input. xi, 82, 86–88, 139, 140, 147, 149–153, 157, 181, 212, 231

LSB Least Significant Bit. 22

MC Marginal Checking. 87, 189, 190

MCD Marginal Checking and Distribution. 87, 186

MDI Manual Data Input. 87, 158, 171, 172

MEW Microwave Early Warning. 16, 17

MI Manual Input. 82, 87, 139, 140, 147, 199, 231

MIT Massachusetts Institute of Technology. ix, 3, 12, 14–16, 36–38, 45, 70, 218

MS Magnetic-Core Storage. 19

MSB Most Significant Bit. 22

MTC Memory Test Computer. 46, 71

OB Output Buffer. 139, 140, 147

OD Other-than-computer-Drum. 140, 146

ONR Office of Naval Research. 13

OTG Optical Timing Generator. 145

OT$_A$ Operate Time A. 119, 121

OT$_B$ Operate Time B. 119, 121

PC Program Counter. 64, 224

PCD Power Control and Distribution. 87, 186

PD Power Distribution. 87

PEC Program Executive Control. 213

PPI Plan Position Indicator. 17, 18, 152, 154, 167

PRRE Photographic Recorder-Reproducer Element. 88, 176, 177

PT Program Time. 119, 121

RD Radar Display. 139, 140, 144, 147, 165, 170, 231

SAB Scientific Advisory Board. 14

SAC Strategic Air Command. 216

SAGE Semi Automatic Ground Environment. xii, 2, 14, 39, 70, 72, 81, 211, 220, 221

SAM Special Add Memory. 224–227

SCC Super Control Center. 78

SD Situation Display. 72, 88, 165, 167–176, 213, 218, 219, 231

SDC System Development Corporation. 125, 209, 211, 212, 222

SDGE Situation Display Generator Element. 87, 157, 170, 171

SDV Slowed Down Video. 56, 149, 153, 157

SOT System Operation Test. 72

SPC SAGE Programming Changes. 212

TD Track Data. 139, 140, 144, 147, 165, 170

TPG Test Pattern Generator. xi, 150, 157, 181, 194

TTY Teletype. 149, 160

TWS Track While Scan. 16

VTVM Vacuum Tube Voltmeter. 113, 115

WD Weapons Director. 168

WSEG Weapon Systems Evaluation Group. 14

XTL Crosstell. xi, 82, 86, 139, 140, 146, 147, 149, 150, 156, 157, 160, 161, 181, 212

Index

1953 Cape Cod System, 71
1954 Cape Cod system, 71

A-1 data system, 217
A9/A10 rocket, 215
ABC, 5
AC supply, 184
accumulator, 10, 27
activate push-button, 158
active computer, 180
adaptation, 212
ADB, 194
ADD, 134, 194
add instruction class, 194
address axis, 160
address register, 131
ADX, 199
AGC, 209
Aiken, Howard Hathaway, 5
Air Force One, 217
air surveillance, 81
air traffic control, 173, 220
alarm, 162
Algebraic System, 65
AM drum, 139
American Airlines, 220
amplidyne, 189
amplitude modulation, 54
AN/FSQ 7, 1
AN/FSQ-32, 222
AN/FSQ-7A, 222
AN/FST-2, 54, 74
analysis
    operational, 211
AND, 100
anode, 8
AOR, 196
Apollo guidance computer, 209
area discriminator, 157, 173

arithmetic element, 25, 27, 82, 119, 131
    left, 119
    right, 119
arithmetic pause, 122
arithmetic unit, 119
Arnold, Henry Harley, 215
ASCC, 5
ASL, 197
ASR, 197
assembly testing, 211, 212
Atanasoff, John Vincent, 5
Atanasoff-Berry Computer, 5
ATC, 173, 220
Automatic Sequence Controlled Calculator, 5
automatic sexadecimal, 53
auxiliary display console, 167
auxiliary drum, 47
azimuth, 153

BCD, 11
Bell System, 217
Berry, Clifford Edward, 5
BFM, 198
BFZ, 198
bias winding, 185
binary coded decimal, 11
binary load card, 207
blip, 81
BLM, 198
block mark, 52
blockhouse, 77
blocking oscillator, 110
blue room, 166
BO1, 160
BO2, 160
BOMARC, 75
BOMARC 1, 160
BOMARC 2, 160

bonus target, 216
Borroughs Corp., 54
BPX, 198
bracket track, 49
branch instruction class, 198
branch on sense, 128
break-in, 129
break-in pulse, 119
breakout, 129
breakout pulse, 119
bright-dim switch, 168
BRM, 198
BSN, 128, 198
buffer drum, 47
Bull Gamma-3, 111
Burroughs, 74
burst, 161
burst time counter, 199
bus, 25
Bush, Vannevar, 3

CAD, 194
calculator mode MC, 189
CAM, 195
camera, 56
Cape Cod system, 71
carry
    end-around, 22
carry look-ahead, 29
cascode, 115
category selection switch, 168
cathode, 8
cathode follower, 10, 92
cathode poisoning, 59
CCA, 209
CD status channel, 146
CD transfer, 140
central table, 214
character compensation plate, 174
Charactron, 39
class code, 126
clock register, 132
closed subroutine, 200
clutter, 17, 154
Clyde, 115
code
    class, 126
    variation, 126
coding, 211
coding phase, 212
coding specification, 211
coil
    peaking, 105
coincident current, 43
Cold War, 1, 216, 221
collector mesh, 175
collector screen, 37
combat center active, 209
command generator, 126
command post, 176
complement
    one's, 21
    two's, 21
COMPOOL, 219
Comprehensive System, 64
computer
    active, 180
computer entry punch, 157
concept verification, 211
conceptual solution, 211
concurrence phase, 211
condition light, 207
conditioned, 101
console, 179
    crosstell, 167
    identification, 167
    intercept control, 167
    surveillance, 167
control element, 25
control grid, 8
control winding, 185
core matrix, 43
core plane, 43
crosstell console, 167
CSU, 195
CSW, 194
cycle
    instruction, 119
    machine, 119
    memory, 119

data link, 71
data-link transmitter, 159
DC inverter, 102

Index   253

DC supply, 185
DCA, 209
DCL, 197
DD, 165
DDA, 7
DDGE, 175
de Florez, Luis, 12
DEC, 46
defense calculator, 70
delay element, 29
delay line, 111
delay line driver, 112
delay unit, 111
Delta Dart, 75
Deltamax, 40
DEP, 196
destructive readout, 42
DFD, 141
digging a well, 36
digit plane driver, 123
digit transfer bus, 25
digital differential analyzer, 7
digital display, 165
digital display generator, 83, 165
digital display generator element, 175
Digital Equipment Corporation, 46
DIM, 195
diode, 8
    vacuum, 8
direct entry section, 158
direct input, 94
direct memory access, 129, 217
direction center active, 209
discrete-constant line, 111
display generator, 165
display system, 165
distributed constant, 111
DMA, 129, 217
documentation phase, 211, 212
Draper, Charles Stark, 72
DRD, 142
drop out, 51
drum, 139
    AM, 139
    auxiliary, 47
    buffer, 47
    field, 140

    LOG, 139, 147, 158, 181
    MIXD, 139, 147, 165, 175, 181
    RD, 139, 165, 181
    status channels, 146
    TD, 139, 165, 181
    timing, 144
drum field driver, 141
drum read driver, 142
drum system, 139
drum, MIXD, 158
DSL, 197
DSR, 197
duplex, 1, 85
duplex maintenance console, 179
duplex switching console, 179, 181
duplex switching facility, 83
DVD, 195

ECCMT, 216
ECH, 196
Eckert, John Adam Presper, 5
electronic counter counter measure technician, 216
emergency change, 212
empty track, 50
end-around carry, 22
Engineering Research Associates, 47
ENIAC, 5
ERA, 47
ERA 1101, 47
erase bar, 145
ETR, 194
executive routine, 201
expansion switch, 168
experimental SAGE subsector, 70, 71
extinction potential, 96

F-106, 75
FAA, 220
Fairchild, 56
fan-out, 93
FCL, 197
FD, 216
feature selection switch, 168
Federal Aviation Administration, 220
Federal Aviation Agency, 220
FETRON, 115

FGD, 74
field, 140
field installation testing, 212
figure of merit, 74
filled track, 50
fine-grain data, 74
fire button, 75
five-digit multiplier, 58
Flexowriter, 53
flip-flop, 10, 103
floating-point, 64
flood gun, 175
FORRESTER, JAY WRIGHT, 13
forward telling, 160
frame, 213
free slot, 152
frequency diversity, 216
FST, 196
full-adder, 132
fully checked design, 218

G/A elapsed time counter, 199
G/A-FD, 159
G/A-TD, 159
G/G, 160
gate tube, 27
gate tube circuit, 101
gating multivibrator, 107
GCI, 13
George, 65
GFI element, 150
gold plated, 215
GOODENOUGH, JOHN BANNISTER, 45
graphical user interface, 218
grid, 8
    control, 8
    screen, 27, 93
    shield, 96
    suppressor, 27, 93, 112
ground control intercept, 13
ground-to-air, 71
Ground-to-Air Frequency-Division, 159
Ground-to-Air Time-Division, 159
Ground-to-Ground, 160
grounded grid amplifier, 98
Group A, 47
Group B, 47

GUI, 218

half-word
    left, 119
    right, 119
HARRINGTON, JOHN V., 54
HARTREE, DOUGLAS, 4
Harvard Mark I, 5
head bar, 141
head crash, 49
heater, 8
height finder request, 160
height-finder radar, 13
high-frequency alternator, 185
high-velocity gun, 37
HLT, 193
HOELZER, HELMUT, 4
horizontal microcode, 33
HORNER, WILLIAM GEORGE, 202
Hughes Aircraft Corporation, 174
Hurricane, 21

I/O selection element, 82
IAS computer, 21
IBM 650, 36, 47
IBM 701, 37, 70, 113
ICBM, 215
identification console, 167
illegal instruction, 194–199
index bits, 126
index interval, 128
index register, 131
induction regulator, 184
inhibit winding, 44
inhibit wire, 44
initial pickup, 74
input
    direct, 94
    pulsed OR, 94
input system, 149
input/output instruction class, 198
instruction
    illegal, 194–199
instruction class
    add, 194
    branch, 198
    input/output, 198

Index   255

    miscellaneous, 193
    multiply, 195
    reset, 199
    shift, 197
    store, 196
instruction control element, 82, 126
instruction cycle, 119
instruction pulse, 119
intercept control console, 167
intercontinental ballistic missile, 215
interrecord gap, 164
inverter, 9, 102
IO address counter, 131
IO control element, 128
IO interlock, 193
IO register, 132
IO word counter, 131
ion repeller mesh, 175
ionizing potential, 96
isolated table, 214

jamming, 216
JFET, 89, 115
jump to subroutine, 201
junction field-effect transistor, 89, 115
Jupiter rocket, 4

KELVIN-HUGHES projector, 176
keyboard control panel message, 158
KILBURN, TOM, 36, 217
kill percentage, 216
KOROLEV, SERGEI PAVLOVICH, 215
Королёв, Сергей Па́влович, 215

LANING, J. HALCOMBE, 65
last in first out, 200
LDB, 194
LDC, 129, 198
least significant bit, 22
leave provision, 201
left arithmetic element, 119
left half-word, 119
LeMay, CURTIS EMERSON, 75
level setter, 98
life curve, 188
LIFO, 200
light gun, 57, 157, 171

light gun amplifier, 172
light pen, 171
Lincoln Laboratory, 14
line, 51
little memory, 208
load center, 183
load unit, 93
LOG, 147
LOG drum, 139, 158, 181
LRI element, 150
LRI monitor, 152
LSB, 22
LSR, 197
LST, 196

machine console, 179
machine cycle, 119
machine specification, 211
machine word, 22
magnetic amplifier, 185
magnetic drum, 47, 139
magnetic tape, 51
magnetic tape adapter, 163
mainentenance consoles, 179
maintenance control element, 82
Manchester Mark I, 217
manual data input element, 157
manual data input unit, 158
manual input, 81, 83
manual mode MC, 190
mapper console, 154
margin of reliability, 188
marginal checking, 60, 188
marginal checking and distribution, 186
Mark I, 5
Massachusetts Institute of Technology, 3, 12
master program, 201
MAUCHLY, JOHN WILLIAM, 5
MC
    calculator mode, 189
    manual mode, 190
    satellite mode, 190
MCD, 186
McGEE, PATRICK, 75
mean time to repair, 7

memory cycle, 119
memory element, 122
memory mapped IO, 17
memory stack, 44
Memory Test Computer, 46
merit, 74
mesh
    collector, 175
    ion repeller, 175
    storage, 175
message, 149, 150
message axis, 160
MEW radar, 16
microcode
    horizontal, 33
microinstruction, 25
microprogram, 25
miscellaneous instruction class, 193
Mischgerät, 4
MIT, 3, 12
MIT Storage Tube, 36
MITRE, 70, 220
MIXD, 147
MIXD drum, 139, 158, 165, 175, 181
mnemonic, 62
model, 211
model change training guide, 211
model maintenance, 212
modem, 54, 217
monostable multivibrator, 107
most significant bit, 22
MSB, 22
MTTR, 7
MUL, 195
multiply instruction class, 195
multivibrator
    gating, 107
    monostable, 107
    one-shot, 107

NAND, 10
NC, 110
NEUMANN, JÁNOS LAJOS, 6
NEWTON, SIR ISAAC, 63
no operation, 208
non-conditioned, 101
nonstandard voltage, 163

NOP, 208
noplex, 85
normalized number, 23
normally closes, 110
north azimuth, 153
north synchronizer, 155
number
    normalized, 23

OD status channel, 146
OD transfer, 140
off centering push-button, 168
Office of Naval Research, 13
OLSEN, KENNETH HARRY, 46
one's complement, 21
one-shot, 107
one-shot multivibrator, 107
ONR, 13
opcode, 32
open subroutine, 200
operand time a, 119
operand time b, 119
operate, 128
operate unit, 128
operational analysis, 211
operational plan, 211
operational specification, 211
operational station, 72
OR
    pulsed, 101
output control element, 159
output storage element, 159
output system, 158

P1 phosphor, 175
P11 phosphor, 165, 176
P14 phosphor, 165
paper tape, 53
    line, 53
parameter testing, 211, 212
party line, 156
pass light gun gate, 172
PCB, 113
PCD, 186
peaking coil, 105
PER, 128, 194
PERT, 66

PETTY, GEORGE BROWN, 219
phase
    coding, 212
    concurrence, 211
    documentation, 211, 212
    post design, 211
    pre-concurrence, 211
    program coating, 212
    program design, 212
    requirements, 211
    test, 212
phase, preliminary design, 211
PHILBRICK, GEORGE ARTHUR, 4
phosphor
    P1, 175
    P11, 165, 176
    P14, 165
photographic recorder-reproducer element, 176
pickup plate, 36
pinch, 89
pluggable unit, 113
Polaroid, 179
PORTER, ARTHUR, 4
post design phase, 211
power control and distribution, 186
power distribution, 186
power reactor, 185
power supply, 183
POWER, THOMAS SARSFIELD, 215
powerhouse, 183
PPI display, 17
pre-concurrence phase, 211
preliminary design phase, 211
printed circuit board, 113
program
    adaptation, 212
    production, 212
program costing phase, 212
program counter, 130, 131
program design phase, 212
program element, 82, 129
program production, 212
program time, 119
programming card, 229
Project Charles, 14
provision
    leave, 201
    return, 201
PRRE, 176
pulse
    break-in, 119
    breakout, 119
    instruction, 119
    timing, 119
pulse amplifier, 93
pulse generator, 109
pulsed OR, 101
pulsed OR input, 94

Q7, 1

R-7 rocket, 215
radar
    height-finder, 13
    MEW, 16
    search, 13
radar mapper, 17, 81
RAJCHMAN, JAN ALEKSANDER, 42
range box, 153
range data, 153
RD drum, 139, 165, 181
RDS, 129, 199
reactance, 90
reactor
    power, 185
    saturable, 185
readout
    destructive, 42
readout pulse, 161
real-time, 13, 219
record, 164
Redstone rocket, 4
register
    address, 131
    index, 131
register driver, 94
relay
    wire contact, 97
relay driver
    thyratron, 96
    vacuum tube, 96
Relay Systems Laboratory, 54
requirements phase, 211

reset instruction class, 199
restoration, 26
return provision, 201
right arithmetic element, 119
right half-word, 119
ripple carry, 134
RISC, 191
ROCHESTER, NATHANIEL, 70
rocket
    A9/A10, 215
    Jupiter, 4
    R-7, 215
    Redstone, 4
ROSSELAND, SVEIN, 4
RSR, 197
RST, 196

S1, 5
S2, 5
SAAICS, 71
SABRE, 220
SAC, 216
SADZAC, 71
SAGE, 2, 14, 69
SAGE Positional Handbooks, 211
SAGE programming change, 212
SAGE System Description, 211
SAH/F, 74
satellite mode MC, 190
SATIN, 220
saturable reactor, 185
scope dope, 165
scope intensification line, 57
screen grid, 27, 93
SD, 165, 167
SDGE, 170
SDR, 128, 199
search radar, 13
SEL, 128, 199
select, 128
select drum, 128
selection element, 128
selection plate, 169
Semi-Automatic Air Intercept Control System, 71
Semi-Automatic Digital Analyzer and Computer, 71

Semi-Automatic Ground Environment, 2, 69
semi-automatic height finder, 74
senior weapons director, 81
sense code, 198
sense unit, 128
sense winding, 43
sense wire, 44
sequence selection program, 201
servo information, 144
sexadecimal, 53
shakedown, 211
shield grid, 96
shift instruction class, 197
shower stall, 46
signal detector, 55
simplex, 82, 85
simplex maintenance console, 179, 181
single-shot, 106
situation display, 165, 167
situation display generator, 83, 165
situation display generator element, 170
sleeping sickness, 59
SLR, 194
software crisis, 219
software engineering, 211, 219
SPC, 212
speaker, 180
special circuits, 112
special side wing, 167
specific design studies, 211
STA, 196
standard level, 91
standard pulse, 91
startover, 83
status channel, 146
storage mesh, 175
storage tube, 36
store instruction class, 196
store-carry gate tube, 29
Strategic Air Command, 216
SUB, 134, 195
subframe, 213
subroutine, 200
    closed, 200
    open, 200

Index 259

substation, 183
Summer Session Computer, 65
Super SAGE, 222
supervisor, 213
supply
    AC, 184
    DC, 185
suppressor grid, 27, 93, 112
surveillance console, 167
switch, 31
switchover, 83
synchronizer north, 155
system evaluation, 211

table
    central, 214
    isolated, 214
TAD, 134, 194
tape
    block mark, 52
    line, 51
    magnetic, 51
    re-record, 52
    read, 52
    record, 52
tape power supply unit, 163
target data, 153
TD drum, 139, 165, 181
TDV, 196
Teledyne Semiconductor, 115
tentative track, 74
test memory, 125
test one bit, 128
test pattern generator, 150, 157
test phase, 212
test register, 132
test two bits, 128
testing
    assembly, 211, 212
    field installation, 212
    parameter, 211, 212
tetrode, 96
thyratron, 96
thyratron relay driver, 96
thyristor, 96
Tic-tac-toe, 220
time counter
    burst, 199
    G/A elapsed, 199
time pulse distributor, 32
time-sharing, 219
timing, 144
timing pulse, 119
timing track, 49
TMU, 136, 195
TOB, 128
toggle switch register, 125
TPG, 150, 157
track, 47
    bracket, 49
    empty, 50
    filled, 50
    friendly, 74
    hostile, 74
    tentative, 74
    timing, 49
transfer
    CD, 140
    OD, 140
transition system, 14
trigger, 96
triode, 8
TSU, 134, 195
TTB, 128
TTY, 160
TTY message, 160
twin add, 134
twin and multiply, 136
twin subtract, 134
two's complement, 21
Typotron, 39, 165, 174

ULSI, 89

vacuum diode, 8
vacuum triode, 8
vacuum tube relay driver, 96
Valley, George Edward, 14
variable inductor transformer, 184
variation code, 126
varistor, 54
Viehe, Frederick W., 42
virtual machine, 64
von Kármán, Theodor, 14

von Neumann, John, 6

War
    Cold, 1, 216, 221
    World II, 1
WarGames, 220
warning light, 162
warning light control element, 162
warning light storage element, 162
warning light system, 162
waterfall model, 211
weapons direction, 81
    team, 81
weapons direction team, 81
weapons director
    senior, 81
Whirlwind, 2, 12, 13, 21
Whirlwind II, 70
Wiesner, Jerome B., 15
Wilkes, Maurice Vincent, 200
Williams, Frederic Calland, 36, 217
Williams-Kilburn-tube, 36
Williams-tube, 36
winding
    bias, 185
    control, 185
wire contact relay, 97
Wolf, William, 66
WOPR, 220
WRT, 129, 199

XAC, 199
XD-1, 70, 191
XD-2, 70
Xenon, 96
XIN, 199
XL, 170
XTL element, 150, 156

YL, 170

Z3, 5
Z4, 5
Zemanek, Heinz, 13
Zener diode, 185
Zener, Clarence Melvin, 185
Zuse, Konrad, 5